商业店面设计 II

（意）斯特凡诺·陶迪利诺／编　张晨／译

辽宁科学技术出版社
·沈阳·

ABOUT THE AUTHOR
作者简介

（意）斯特凡诺·陶迪利诺

斯特凡诺·陶迪利诺设计公司 创意总监

Stefano Tordiglione
Creative Director
Stefano Tordiglione Design Ltd

Stefano was born in Napoli, Italy and he studied and worked in New York and London for over ten years. His first design experience started in London in 1991. Since then he has worked for internationally acclaimed Italian studios specialized in retail, luxury hotel and resorts, residential development and private yachts.

He is also an artist and his works are part of international private contemporary art collections. It's also worthy to mention the fact that he worked for UNICEF as Project and Art Director in several international events organized in Italy. His numerous artistic experiences have made him develop excellent intuition into aesthetics and design.

The design team at Stefano Tordiglione Design Ltd. has a strong international background with architects and designer. They have directed many commercial projects of the studio including Brooks Brothers Fashion Boutique, Dinh Van Jewellery boutique, Wellendorff jewellery boutique, apple & pie Child-shoe Boutique and Sal Curioso Spanish Restaurant etc.

Recent Awards:
2014 Golden Bund Awards: Certificate of Excellence of Best Commercial Design
2013 Perspective Awards: Certificate of Excellence of Restaurant – Interior design
2013 Asia Pacific Interior Design Awards: Best 10 Food Space
2013 Asia Pacific Interior Design Awards: Best 10 Shopping Space
2013 Global Design Awards: Certificate of Merit (Hospitality & Entertainment)

Stefano 在意大利的拿波里出生，并在纽约和伦敦留学和工作超过十年。他的设计经验可以追溯到1991年的伦敦。自此开始，Stefano 在享誉国际的意大利公司工作，擅长于高端商铺、酒店和会所、私人住所和游艇的设计。

他原来是一位艺术家，其创作是现代国际艺术私人收藏系列的重要组成部分。不得不提的是，Stefano 曾于联合国儿童基金会担任项目与艺术总监一职，在意大利统筹了不少大型的国际性活动。Stefano 的丰富艺术经验令他建立了独特的美感与设计触觉。

Stefano Tordiglione Design Ltd的设计团队由拥有丰富国际经验的建筑师和设计师组成，他们承担了一系列商业项目的设计，包括 Brooks Brother 专卖店、Dinh Van 珠宝店、Wellendorff 珠宝店、apple & pie 童鞋专卖店和 Sal Curioso 西班牙餐厅等。

近期奖项：
2014 金外滩奖：最佳商业设计 – 优秀奖
2013 透视大奖：最佳餐厅设计 – 室内设计
2013 亚太区室内设计大奖 – 最佳10名（餐饮空间）
2013 亚太区室内设计大奖 – 最佳10名（购物空间）
2013 环球设计大奖：优异奖（消闲及娱乐场所）

CONTENTS 目录

Foreword 前言	004
Clothing and Accessory Store 服装、配饰	006
Restaurant 餐厅	076
Café and Bar 咖啡厅、酒吧	138
Grocery and Bakery 食品、烘焙	178
Modern Living 居家生活	218
Cultural Facility 文化	272
Index 设计者（设计公司）索引	278

World Shopfront Design: DNA of Brand Building

Shopfront design or façade design is getting more attention from brands nowadays. It is the 'skin' of the brand's DNA, should it be contemporary or traditional, sophisticated or casual. A simple glance of the shopfront tells us about the core values of a brand. In a business world that all brands are competing with each other, it is natural that brands are putting more effort on shopfront design to make itself stand out from the crowd.

Shopfront usually consists of a few consistent elements: the logo, the entrance, the window and the structure links up all these elements together. While the logo mainly belongs to the scope of the visual identity, the rest of the structure belongs to the work and imagination of the interior designer and architect. The position of the entrance is the first element to consider in the design process as the entrance will dictate the traffic and the flow of circulation. It has direct interaction with the customers, although most of the time it is not very obvious. Served mainly as a passageway, if handled well it creates the first contact with customers with fresh and new sensations. The window is the attraction of a brand. Intrinsically linked with the visual merchandising strategy, the window provides the first display opportunity of the products. It highlights the 'cream' of the collection and attracts customers to come in the shop. Located mostly at eye level or slightly higher than the height of our body, it is a human-scaled space that we can connect with, leaving an imaginary space for us to visualize the fantasy about how we will look like after we have put on the product. As for the structure that links up all of the above pieces, it occupies most of the space in the façade. Its massive scale makes it stand out from the surrounding environment, sending a strong and bold message to all passerby 'I am here!' Depending on the structural limitation of the building, all of these elements jointly form the design of the shopfront to tell a unique story of the brand before we have entered the boutique.

As an integral part of the shop image design, it is important to have the shopfront design link up with the interior of the boutique. The exterior and interior are like two sides of the same coin, while they may manifest different expression, they form the same brand and some elements should be consistent inside out. It is easier said than done since in most of the cases the shopfront design has to take in consideration of the limitation of the structure of the building while the interior can be changed relatively more flexibly. Many structural and engineering regulations pose an impact on the façade design, limiting and shaping the creativity of designer and architect. Paradoxically, as long as the interior does not affect the structure of the building, designers have carte blanche to unleash their dream. These are two interesting aspects in the design process that only designer and architect who incorporate both area - façade and interior design - under their scope of work can resonate with. The joys and tears in the process of creation and realization are valuable experiences in the career of designer and architect.

A new trend in the shop front design is the stress on the creation of identifiable motif or symbols that customers can easily recognize the brand. These motif can be the logo - which has to be handled well otherwise it will become too conspicuous - or can be a common pattern of the brand. All of these elements can reinforce the visual image and hence the identity. We have been designing for the American fashion brand Brooks Brothers for the third boutique now. We are particularly proud of the creation of the fleece on the grid pattern of its façade design. The fleece has been the icon of the brand since 1850 signifying its roots in fine woolen products. The grid pattern is reminiscent of a classic window pane design of a 20th century mansion situated in New York. Together they form a well blended decoration on the façade to symbolize its heritage, quality and service. This is a gentle message to all who pass by the boutique. Not everyone will know the brand of Brooks Brothers immediately, however, people passing by the boutique will notice this special ornament and associate this with the brand. The visual impact on customers is gradually taking place, forming a subtle stage of brand education. This progressive manner of brand building is intrinsic to the values of the brand.

An eye catching shop front is the basic necessity for brands due to its strong linkage to sales performance. As the façade is the direct message of the service and products of the brand inside, it builds up the anticipation of customers before entering the boutique. Shop fronts sends out the welcoming message and encourage the impulsive buying behaviour. Many psychologists have researched about the shopping behaviour of people and the co-relation with the effect of an attractive window. This area is worth further studies to enhance the power of a successful shop front design. At the end of the day, an important function of façade design is to encourage sales. As commercial as it sounds, this is the key purpose from the client's perspective which we must address.

In short, a facade is a mask and the mask becomes the identity of the brand. It is a game of attraction and we need to create a desire of being and wanting to make a statement. This mask in general is to cover the appearance and have a new image. Similarly, the facade is the voice of the brand and the communication to the outside world. The final and ultimate goal of designer and architect is to unleash the imagination and maximize the practicality to create the 'mask'. This is an ultimate challenge to become the most attractive brand among all.

全球店面设计：品牌的建筑 DNA

如今的品牌越来越重视店面设计和外墙设计。无论呈现何种风格，现代或传统，复杂或随性，店面设计都是品牌 DNA 的"脸面"。店面会通过给人的第一印象传递品牌的核心价值。面临商业世界中的激烈竞争，品牌自然将更多的精力投入店面设计，力争在竞争者中脱颖而出。

店面通常由一些固定元素组成：标识、正门、橱窗和连接所有元素的整体结构。标识主要属于视觉形象的范畴，其余结构则属于室内设计和建筑设计的领域。正门的位置是设计中首先要考虑的问题，因为它会直接决定人员流动方向和模式。尽管大部分时间并不非常明显，正门的位置与顾客之间存在着直接互动。作为店面的主要通道，设计合理的正门会为顾客带来新鲜、惊喜的第一感觉。橱窗是品牌的吸引力所在，它与视觉营销策略存在内在联系，为产品提供第一手的展示机会。橱窗突出的是产品中最精华的部分，吸引顾客移步店中。橱窗的位置一般与视线高度相同或略高于身高，是一处互动空间，可以激发顾客发挥想象，幻想自己使用了产品以后的美妙效果。至于连接以上所有元素的结构，则占据了店面外墙的大部分空间，显著的风格从周围环境中脱颖而出，向路过行人传递强烈而直白的信号"我在这里！"在不同的建筑结构限制下，所有这些店面元素共同构成店面的设计，使得顾客在进店之前便可以了解到独特的品牌故事。

作为店铺形象设计中不可分割的一部分，将店面设计与室内设计紧密联系十分重要。外墙和内墙像一枚硬币的两面，即便可能展示不同的风格，却会共同构成一个品牌的品牌形象，室内外设计中还会使用到一些重复元素，保持一致。可这终究是说起来容易，做起来难，因为大多数情况下，店面设计必须考虑到建筑本身的结构限制，而室内设计就相对灵活一些，易于改变调整。许多建筑和工程规定都对店面设计构成影响，也对设计师和建筑师的创作过程构成限制和影响。与此矛盾的是，只要室内设计不对建筑结构产生影响，设计师就可以尽情发挥、设计。设计过程中的这两个有趣现象是只有同时涉及外墙和室内设计领域的设计师和建筑师才能体会得到。创意和施工过程中的快乐和苦涩都是设计师和建筑师的职业经历中的宝贵经验。

简而言之，店面外墙是一张面具，这张面具就是品牌的身份和形象。店面设计是一个关乎吸引的游戏，设计师需要创造出期待彰显个性的欲望。总的来说，这张面具会为店铺带来新的形象。外墙的效果与此类似，是品牌的形象代表，负责与外界沟通。设计师和建筑师的终极目标则是在"面具"的设计中解放想象力，强化实用性。而这也是使品牌变得出众而有吸引力的终极挑战。

斯特凡诺·陶迪利诺 (Stefano Tordiglione)

JUN AOKI & ASSOCIATES

LOUIS VUITTON MATSUYA GINZA RENEWAL

Ginza, Tokyo, Japan

路易威登松屋银座店翻新项目

青木淳建筑设计事务所 / 日本，东京，银座

The new façade of Louis Vuitton Matsuya Ginza is inspired by the history of Ginza, the district that used to be known for its art deco design. Ginza was the entrance of Tokyo, adjacent to Shimbashi, from which the very first railway station of Japan stretched to the port and led to the foreign Country. The 'modern' atmosphere the forefront Ginza acquired derived from art deco patterns in relation to *edo-komon*, the pattern of traditional Tokyo and the highly abstract and stylized geometric pattern in repetition.

Based on Louis Vuitton's *damier*, which also is a repeated geometric pattern, the façade of Louis Vuitton Matsuya Ginza becomes a softer version of *damier*, imbued with delicacy and richness that is found in organism.

From *edo-komon* to art deco. Art deco to the soft *damier*. This is a journey to the history of Ginza.

Gentle bulges and dents elaborate the façade of opal beige reliefs. With these patterns, the façade reveals various appearances in sunlight, and also during the night, the LED lights behind the reliefs light up the façade to render another expression reminiscent of Louis Vuitton's *monogram*.

Renovated Area: 1,475m² (projected elevation)　Construction Period: Jan 2013 – Sep 2013

General Specification:

- **Exterior wall**
  2-8F: Aluminum panel, pearl paint
  Casting/Spinning/Cutting sheet assembling, LED lights
  1F: Limestone, honed finish/sandblast finish
- **Entrance**
  Marble, polished curve finish/Honed finish
  Super Clear Glass t12+12
- **Windows**
  Metal frame: Stainless steel, Cross HL, Bronze colour coating
  Glass: Super clear glass t12+12mm

通用规范：

- 外墙
  2-8 楼
  铝板，珍珠涂料
  铸造 / 旋转 / 板材切割和组装，
  LED 照明
  1 楼
  石灰石，亚光表面 / 喷砂表面
- 入口
  大理石、曲线抛光表面 / 亚光表面
  超级透明玻璃 t12+12
- 窗
  金属框架
  不锈钢，Cross HL，青铜色涂层
  玻璃
  超级透明玻璃 t12+12

路易威登松屋银座店的新外墙设计灵感源自银座的历史，这里曾因为其艺术装饰风格而被人们所熟知。银座毗邻新桥，是东京的入口，日本最早的铁路火车从这里出发到达港口，走向世界。结合了传统东京风格图案"江户小纹"和高度抽象且风格化的重复几何图案的艺术装饰风格赋予银座前沿洋溢的"现代"氛围。路易威登松屋银座店的外墙设计使用了路易威登经典的棋盘格图样，棋盘格也是一种重复的几何图案，而这个店面成为了我们首个弱化版的棋盘格设计，充满有机体的精致感和丰富感。

从"江户小纹"到摩登设计风格。再从摩登设计风格到弱化的棋盘格设计。这里展现出的是银座的历史变迁。

温柔的凸起和凹陷构成米白色的浮雕式墙面。外墙上的这些图案在日间和夜间展现出不同的模样。夜幕降临时，浮雕背后的LED灯亮起，使外观变得十分不同，让人联想到路易威登品牌的字母组合。

面积：1475平方米（突出立面） 施工时间：2013年1月-9月

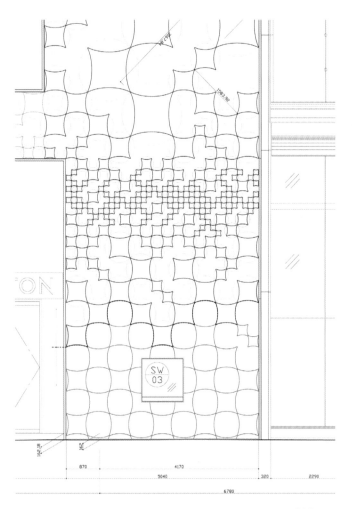

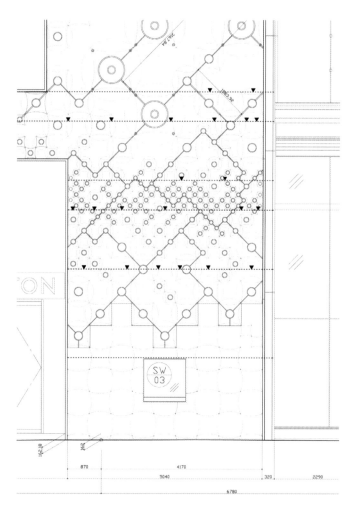

BARBARITOBANCEL ARCHITECTES

DIOR MIAMI FACADE

Miami, USA

迈阿密迪奥

barbaritobancel 建筑师事务所 / 美国，迈阿密

Through large curved movements of white concrete, clear figures of the 'plissée' take shape, between which the spaces of the boutique slide in. The nobility of the smooth and delicate surfaces is given by a contemporary material made of ultra high-density concrete and by marble powder.

The project also respects its 'commercial' nature. The building does not want to appear as an institution or a museum; and its generous shop windows open to the immediate public place. The drawing is influenced by the suggestive inspirations of Miami, images of sun and beaches along with an idea of dynamism, youth, and contemporary design. Architecture and Haute Couture unite in the common desire to seduce. The elegant white UHPC (ultra high performance concrete) was designed by BarbaritoBancel.

Completion Date: March 2016 Photographer: Alessandra Chemollo

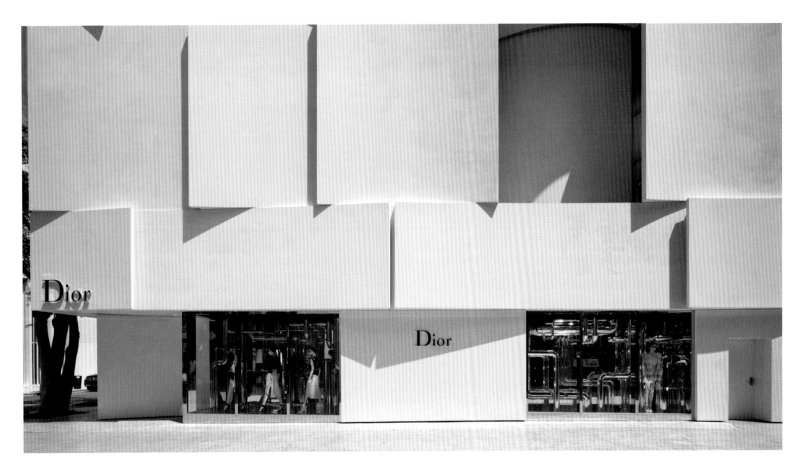

011 • 服装、配饰

白色混凝土的曲线造型打造出的"泡泡纱"效果将精品店空间包容在内。超高密度混凝土和大理石粉末等现代建筑材料打造出流畅而精致的表面。

设计方案还充分考虑到店面的"商业"性质。店面不应该是一个严肃的博物馆,而是向公共空间直接开放的通透橱窗。设计灵感来自迈阿密的骄阳和沙滩,以及活力、青春和现代元素。建筑与时装结合在一起,共同散发诱人的魅力。优雅的白色UHPC(超高性能混凝土)也是由BarbaritoBancel建筑师事务所设计打造的。

建成时间:2016年3月　摄影:亚历山德拉·切莫洛

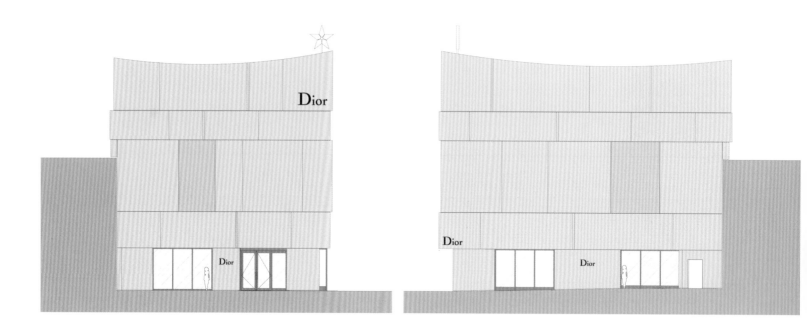

HANGZHOU GUAN INTERIOR DESIGN

G'RSAGA MEN'S WEAR

Ningbo, China

G'RSAGA（浪琴海）品牌男装

杭州观堂设计 / 中国，宁波

G'RSAGA Men's Wear sticks to the philosophy of style and trends, pursuing the integration of classics and fashion and demonstrating a contemporary lifestyle of elegance, sexiness and decency.

The client wished to combine a few vintage elements in its 2014 brand identity and create a simple light industry concept.

Therefore the designer chose wood as the core texture. Blue wooden grain was used with minimalist logo, while the interior employs wooden flooring, desktop, ceiling together with black counter, division and decorative lines, generating a natural, relaxing atmosphere.

Designer: Zhang Jian Project Features: Commercial space, branded clothing Completion Date: April, 2014 Area: 140m² Materials: Wood, wrought iron Photography: Liu Yujie Text: Tang Tang

G'RSAGA（浪琴海）品牌男装，坚守潮流主义，追求经典与时尚的融合，优雅、性感、得体，致力于诠释当代青年的生活思维方式。

2014年度的形象设计，客户希望融合些许复古元素，整体呈现简约的轻工业风。

因此在设计手法上，设计师围绕木质奢华展开，材质上选择以木质为主，店面外观选择蓝色木纹配以品牌简洁的标识，同时在内部装饰上打造了木质地面、木质桌面、木质顶面等，配以黑色货柜、隔断、线条，带给客人放松、自然的氛围。

设计者：张健 空间性质：商业空间，品牌服装 完工时间：2014年4月 面积：140平方米 主要材料：木质，铁艺 摄影：刘宇杰 撰文：汤汤

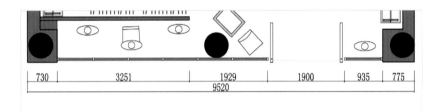

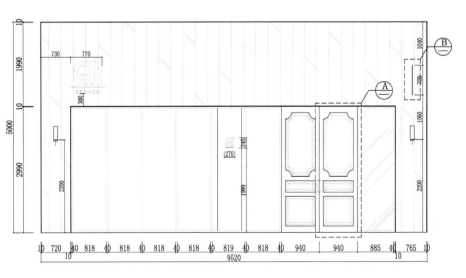

1:50

GOSPLAN ARCHITECTS

'IL SALOTTO'

Genova, Italy

"起居室"

gosplan 建筑师事务所 / 意大利，热那亚

When Sara Busiri Vici and Matteo Brizio came to gosplan office looking for someone who could design their clothes shop, they brought two requests: the brand had to be 'il salotto' (Italian for 'the living room') and the site had to be delivered within three months.

The boutique is situated in the historical center of Genoa, at the ground floor of an ancient building. The main room has vaulted ceilings and large openings to the narrow streets of the medieval tissue, where the sun barely shines. The entrance is marked by a marble gate that dates from XV century and which probably was the main entrance to the whole building in the past. The iron door is a free interpretation of doors of ancient Genoa palaces. It is made of perforated metal sheets. The small holes are a 'metaphore' for the large ancient nails, while the large hole in the center replaces the door knocker. The circular hole also focuses on the boutique logo and it is aligned with the two circular medallions carved in marble pillars alongside of it. The iron door enhances the ancient marble gate that previously was harmed by a trivial rolling shutter.

Area: 80m² Photographer: Anna Positano

当莎拉·布西里·维奇和马泰奥·布里齐奥来到 gosplan 建筑师事务所希望寻找到设计师，为他们的服装店打造店面，他们提出了两个要求：品牌需要具有"起居室"一般的氛围，场地施工需要在三个月内完成。

这间精品店位于历史悠久的热那亚市市中心，占据了一座古老建筑的一层空间。其中主要房间的天花板呈拱形，宽大的窗口面向街道。这些街道仍保留着中世纪元素，阳光几乎照不到这里。店铺正门的大理石大门可以追溯到15世纪，这扇大门在过去很可能也是整个建筑的入口。铁门是对古代热那亚宫殿大门的自由诠释，它由多孔金属板制成，板材上的小孔象征着巨大的古代钉子，中央位置的大孔则代表门环。圆孔也体现了精品店的品牌标识，并与刻在两侧大理石柱子上的两个圆形徽章相互呼应。铁门突出了历史感厚重的大理石大门结构。

面积：80平方米 摄影：安娜·波西塔诺

BBC ARQUITECTOS

LOCAL JUNÍN

Buenos Aires, Argentina

胡宁商业空间改造

BBC建筑师事务所 / 阿根廷，布宜诺斯艾利斯

The commission consists of the enhancement of an old neoclassical cinema city of Junin, Buenos Aires Province, changing its use to a commercial space for clothing.

Considering the condition of historic heritage, the basic premise was giving its characteristic style that was disjointed in above modifications.

The project seeks to achieve the reinterpretation and a contemporary architectural language through different operations.

The first is the redefinition of the facade by determining a continuous plane with a lower part glazed and the defacing on its crowning. The second step consists in the qualification of the access bounded by the lower glass facade and the side pillars virtualization. A metallic prism hosts the display window opposing its own weight against the access space.

The transformation is completed with a total release of the ground floor with side supports in one side and an elevation of the ceiling to 5.5 meters, achieving an open floor 30 meters long and 12 meters wide with great flexibility.

Designers: Angela Bielsa, Luciana Breide, Manuel Ciarlotti Bidinost
Completion Date: 2014 Area: 530m² in renovation Photography: Manuel Ciarlotti

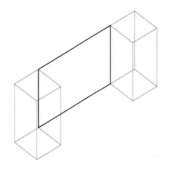 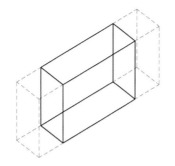 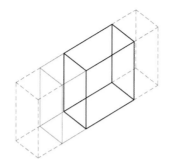

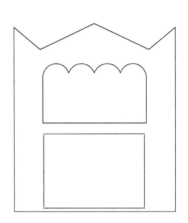 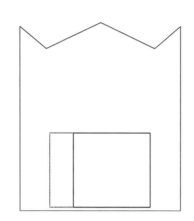 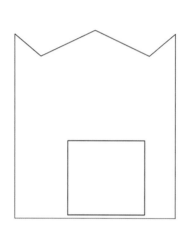

项目中的委托内容是将布宜诺斯艾利斯省古老的新古典主义风格胡宁电影城改造成出售服装的商业空间。

考虑到历史建筑的状况，项目的基本前提是赋予前文指出的标志性风格特征。

项目需要通过不同的操作实现建筑的重新诠释和现代建筑语言的表达。第一步是外墙的重新设计，用一面较低的玻璃墙面配合凸面加工形成一个连续墙体平面。第二步是利用较低的玻璃墙面对入口加以限制，完成两侧柱子的虚拟化设计。金属棱镜具备展示窗的功能，与入口空间形成对比感。

改造工程的最后一步是一楼的整体释放，侧面是支撑结构，天花板高度5.5米，形成30米长，12米宽，具备极高灵活度的开阔楼层空间。

设计师：安吉拉·贝尔萨，卢恰娜·布雷德，曼努埃尔·希亚洛蒂·比迪诺斯特　完成时间：２０１４年　面积：改造５３０平方米　摄影：曼努埃尔·希亚洛蒂

BBC ARQUITECTOS

LOCALES ROCA

Gral Roca, Provincia de Rio Negro, Argentina

罗卡将军镇商业空间改造
BBC建筑师事务所 / 阿根廷，里奥内格罗省，罗卡将军镇

The order stipulates the construction of commercial properties which are located in the central area of General Roca, Rio Negro.

Three commercial offices are determined with different hierarchy, where two of them are the subsidiaries smaller offices, should be indeterminate without knowing the program or future uses.

A rectangular and articulate body is constructed to get space sequences in the main room through the courtyards and supporting framings and skylights projections. Both topics regulate the travel and qualify the transversal spaces, which is demarcated by the separation between supports and enclosures. The rear courtyard space will take deposits and changers off releasing diaphanous perspective.

The façade of the building appears as a single asymmetric compact piece. It consists of an independent frame lined with sheet and glass, able to respond to different specifications and requirements of each location.

Designers: Angela Bielsa, Luciana Breide, Manuel Ciarlotti Bidinost   Area: 600m²   Photographer: Manuel Ciarlotti

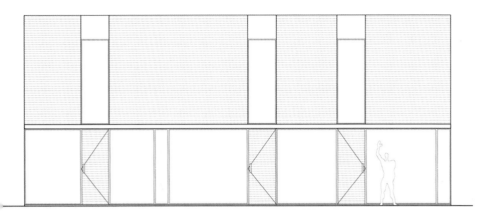

本案是位于里奥内格罗省的罗卡将军镇中心一处商业地产的建筑工程。三个商业办公室之间呈现不同的层次结构，其中两个是次级办公室，在使用目的和未来功能不能确定的情况下应保留设计空间。

设计师建造了长方形，富有表现力的机体，通过庭院和支撑框架及天窗的安排体现主房间内的空间顺序。通行路线由两个主题，限制横向空间。支撑结构和包围结构的分离划定了横向空间。储物和更衣空间安置在后院，打造出精致的整体视觉效果。

建筑外墙呈现不对称的紧凑结构。它由一个独立的框架构成，配合衬板和玻璃，能够满足每个位置上的不同规格和要求。

设计师：安吉拉·贝尔萨，卢恰娜·布雷德，曼努埃尔·希亚洛蒂·比迪诺斯特　面积：600平方米　摄影：曼努埃尔·希亚洛蒂

6A ARCHITECTS
PAUL SMITH

11 Albemarle Street, London, UK

保罗·史密斯

6a 建筑师事务所 / 英国，伦敦，阿尔伯马尔街 11 号

The new Albemarle Street shop front for Paul Smith builds on a familiar material tradition in London. Cast iron forms an understated background to the city's streets; its railings, gratings, balconies, and lamp posts. Paul's brief was an eclectic collection of references, images, textures and traditions, encompassing military medals, woven hats and finely drawn gold ingots alongside sharp tailoring, the soft fall of cloth, craftsmanship and delight in surprise. How these disparate influences might find the restraint needed to engage with the neighbourhood at the same time as contributing to its future identity was key, in doing so offering a rare resistance to the increasingly homogenous global high street.

The ground floor rustication of Georgian townhouses and the ornamental language of 18th century shop fronts were reinterpreted and abstracted in a sinuous pattern of interlocking circles cast into a new solid iron façade. The repetition of the typical Regency shape brought an optical complexity which, with the play of sunlight and shadow, turns the pattern into a deep surface texture. Seen obliquely it seems woven, like a fine cloth.

The surface is further enlivened by the latent makers' marks of the casting process and the natural patination of the cast iron. A more intimate discovery is to be made in the trio of small drawings by Paul cast directly into panels scattered across the façade.

Curved windows project from the darkly textured iron as luminous vitrines, with a nod to the curved glass of the nearby arcades. A secret door of stained oak lies flush with the cast iron panels: the inverted carving of the timber recalls the mould and sand bed prepared for the molten metal.

The cast iron panels curve in to the recessed oak entrance door, a gently bowed iron step evokes worn away treads. Over time, the iron threshold will polish under foot, recording the life of the building in its material.

Project Team: Tom Emerson, Stephanie Macdonald, John Ross, Owen Watson, Noelia Pickard-Garcia, Johan Dehlin
Photographer: 6a Architects/ DG: David Grandorge

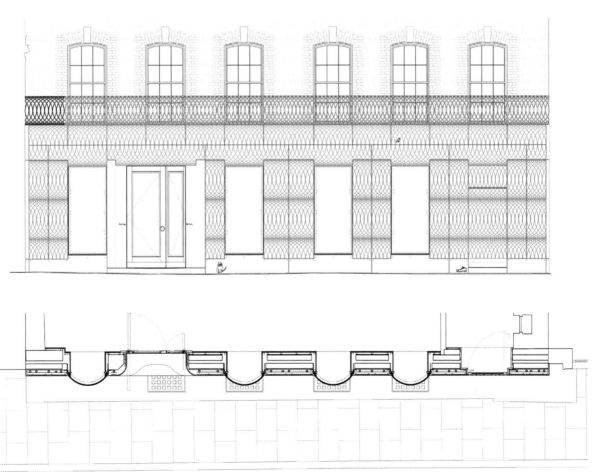

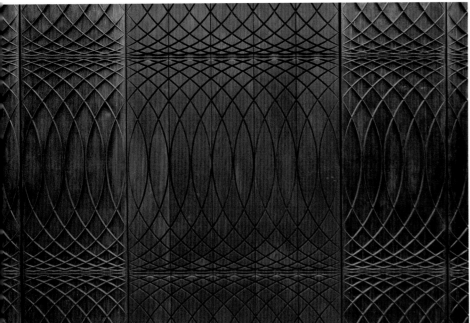

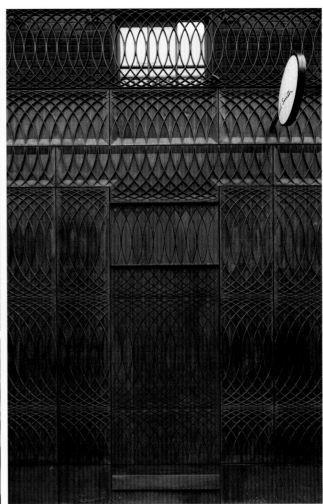

保罗·史密斯品牌在阿尔伯马尔街的新店采用了伦敦常见的传统建筑材料。铸铁元素构成城市街景的朴实背景：栏杆、格栅、阳台和灯柱。委托方的项目要求集合了形形色色的参考文献、图片、材料和传统习惯，包含了军事奖章、编制帽子和绘制精美的金条等元素。设计中的关键是这些不相干的元素如何与周边环境产生互动，同时如何塑造未来形象，抵御全球商业街同质化的趋势。

设计师对乔治亚式排屋一楼的朴实感和 18 世纪店面的装饰语言进行了重新解读，用错综复杂的抽象连环圆圈图案打造出新的铁艺外墙。典型摄政王朝样的重复增加了设计的复杂度，玩转阳光和阴影，将图案刻画成深邃的表面纹理。斜向观察，整个表面看起来就像一块精致的布料。

制造商的标识使得这个平面更加具有活力，设计师决定将相同图案的面板推广应用到整个外墙，尝试更进一步的探索。

弧形窗户从深色的铁框架结构中探出，染色橡木密门与铸铁面板齐平，反向雕刻的木料让人想起熔融金属过程中用到的模具和沙床。

铸铁面板的弧线造型连接隐藏的橡木大门，弧度柔和的铁台阶让人想起破旧的阶梯。随着时光流逝，铁门坎将在脚下磨得光亮，以材料的形式记录下建筑的生命历程。

项目团队：汤姆·爱默生，斯蒂芬妮·麦克唐纳，约翰·罗斯，欧文·沃森，诺莉娅·皮卡德加西亚，约翰·德林 摄影：6a 建筑师事务所 / 大卫·格兰德奥热

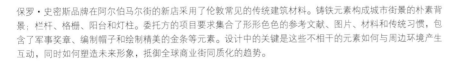

CLAUDIO SILVESTRIN ARCHITECTS

GIADA STORE BEIJING

Beijing, China

北京迦达旗舰店

克劳迪欧·席维斯金建筑师事务所 / 中国，北京

Distinctive architect Claudio Silvestrin, designer of the Giada Milan store in Montenapoleone, has created a contemporary and timeless architecture, where ancient and modern embrace one another in a calm and elegant space, giving a luxurious feel and full expression to the Giada collection. The 250m^2 Giada flagship store in Beijing is facing the prestigious Jinbao Street, with entrances from the street as well as the Jinbao Place shopping mall.

Comprised of truly natural materials and elements: life-giving water, stone walls and floors, cast bronze and natural leather, full height doors and mirrors-all united in harmony through a linear and decisive geometry, rich in sophisticated details. A travertine water trough runs along the entrance wall. Blocks of cast bronze provide pedestals for the products, vitrines and a screen for the cash desk.

The rough surface of raw limestone totems create a textural backdrop to the soft clothes, which are hung on square frame rails with a bronze finish. The design of the store's exterior is revolutionary: an enormous façade made of yellowish porphyry stone, 15 meters wide and 9 meters tall, with a singular, smooth, bronze finish door adjacent to an equally proportionate clear glass window looking into the boutique itself. The façade is instantly striking. It is characterised by huge, illuminated cracks breaking the surface of the seemingly indestructible stone.

The idea behind this appearance is a poetic metaphor representing the rising energy and fire of a dragon. In the architect Claudio Silvestrin's own words:
'A young dragon dwells underground
Unknown
He dreams of seeing the sky
He pushes upwards
In rising, the earth's surface begins to crumble
Through the cracks
Light emerges'

Designer: Claudio Silvestrin   Area: 250m^2

蒙特拿破仑大街上迦达米兰店的设计师，著名建筑师克劳迪欧·席维斯金打造出一个现代而经典的建筑，古代和现代在冷静而优雅的氛围中交融，赋予迦达的品牌产品奢华感和表现力。

占地250平方米的北京迦达旗舰店门口就是著名的金宝街，从金宝街和金宝购物中心都可以进入店内。店面设计采用真正的天然材料与元素：生命之水、石头墙壁和地面、锻造青铜和天然皮革、落地门窗，所有元素都和谐地以流线与几何形式结合在一起，细节复杂而丰富。

一条洞石水槽贯穿整个入口墙壁，锻造青铜块为品牌产品提供展示基座，橱窗和屏风组成收银台。

未经加工的石灰岩图腾粗糙表面为柔软的服装创造了极具质感的背景，服装展示使用的是青铜色表面的方形框架轨道。

店面的外观设计极具革命性：由淡黄色斑岩石材组成的巨大立面宽15米，高9米，墙上一扇光滑的青铜色门靠近相同大小的透明玻璃窗，充分展示店内空间。

店面外墙效果引人注目。巨大的看似坚不可摧的石材表面上条条裂缝透出神秘的光亮。

设计背后的理念来自一项充满诗意的比喻，代表上升的能量和龙之火。引用建筑师克劳迪奥·席维斯金的话："幼龙盘踞地下／尚且不知／自己看见天空的梦想／他向上推力／上升的过程中，地面开始瓦解／透过裂缝／出现了光。"

设计师：克劳迪欧·席维斯金　面积：250平方米

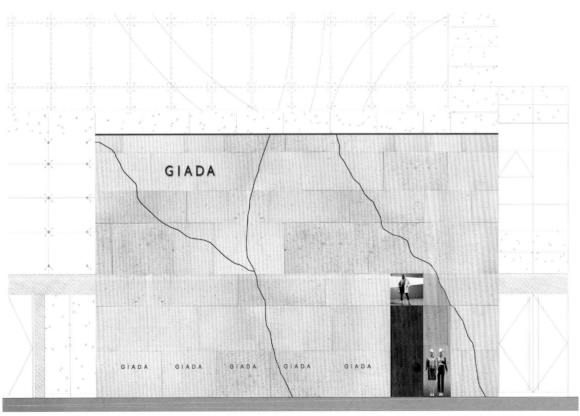

DPJ & PARTNERS ARCHITECTURE

CHRISTIAN DIOR FLAGSHIP

464, Apgujeong-ro, Gangnam-gu, Seoul, Korea

克里斯汀·迪奥韩国首尔旗舰店

DPJ & Partners 建筑师事务所 / 韩国首尔江南区狎鸥亭 464 号

The client wanted the building to represent Dior and to reflect Christian Dior's work. So they wanted the surfaces to flow, like the couturier's soft, woven white cotton fabric. These surfaces, which soar into the sky and undulate as if in motion, crossed by a few lines, are made from long moulded fiber glass shells, fitted together with aircraft precision.

In Seoul, where the quadrangular buildings align with the avenue, and which are all occupied by leading international fashion labels, the building stands out like a large sculptural tribute to Dior, inviting everyone to step inside.

The entrance, where two shells come together, is a sort of modern lancet arch, in which two metal mesh surfaces cross in line with the clothing metaphor. Once inside, the customer makes a succession of discoveries – a feature typical of the interiors designed by Peter Marino.

Construction Period: June 2013-June 2015 Main Building Elements: Glass Fibre Reinforced Plastic (GFRP) façade composed of self-standing 11 panels of 6m×20m, anodized aluminum panel claddings.

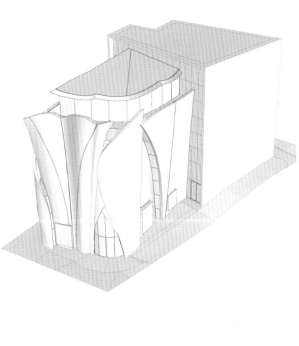

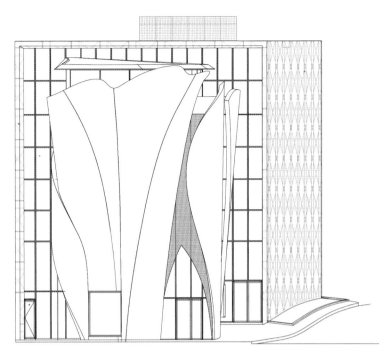

委托人希望建筑设计能代表迪奥品牌精神，反映迪奥的工作成果。他们希望店面的外观具备流线感，就像柔软的白色纯棉纺织面料一样。这些表面朝向天空，呈现出波浪的形状，仿佛在律动，中间的几段线条由模压玻璃纤维外壳制成，安装过程拥有航天设备一般的精准。

首尔的四边形建筑与街道平行，沿街的都是引领潮流的国际时尚品牌。迪奥的店面设计像一座巨大的雕塑，彰显品牌风采，引人进店一探究竟。

正门处的两个框架结构连接到一起，呈现类似现代挑尖拱的形状。这里两片金属网面交叉，象征服装。进入店面室内，有一连串的惊喜发现在等着顾客——这是彼得·马里诺室内空间设计的典型特征。

施工时间：2013年6月–2015年6月　主要建筑元素：11个独立的6米x20米面板组成的玻璃纤维强化塑料（GFRP）外墙，阳极电镀铝覆面面板

UUFIE

PORTS 1961 SHANGHAI

Shanghai, China

宝姿 1961 上海

UUfie／中国，上海

Located at a major high-end commercial district at the intersection of Changde Road and Nanjing West Road in Shanghai, a new façade is created for the fashion house Ports 1961's flagship store. The facade is representative of the future vision of Ports 1961 that brings together its origin and evolution. The design evokes the idea of a landform that resembles an iceberg floating freely in the ocean; the building having a sense of being undulated, expanding and contracting, as if shaped by its environment. The facade demonstrates the possibilities of design experimentation, showing the transformation of form, material and technology, while still bringing aspect of both traditional and contemporary interpretations. The structural grandeur of the building attracts attention while its ambivalent nature is uniquely changing with its surroundings.

The facade is composed of two types of glass blocks; a standard 300mm x 300mm glass block and a custom 300mm x 300mm x 300mm corner glass block. Unlike most glass block masonry constructions, it is not limited to one vertical plane. The combinations of the two types of glass block create a sculpted three-dimensional façade exhibiting cantilevered structures. By incorporating innovative structural engineering and inventing a new joining system in the block itself, an elaborate ornamental stepping canopy is achieved that naturally angles to the flow of pedestrian and allows for four bow windows to be visible from all directions.

The usage of satin glass block and shot blasted stainless steel material is in great contrast to the chaotic city. During the day, it mutes the surroundings, while subtly reflecting the sunlight. In the evening, the view is icy and crisp, and the surface illuminates with embedded LED lights integrated into the joints of the masonry. The differing geometries and changing perspectives of the facade expresses the transformative nature of the city and the people of Shanghai.

Façade Contractor: J. Gartner & Co. (HK) Ltd.   Area: 1,145m²   Construction Period: 2014 - 2015   Photographer: Shengliang Su

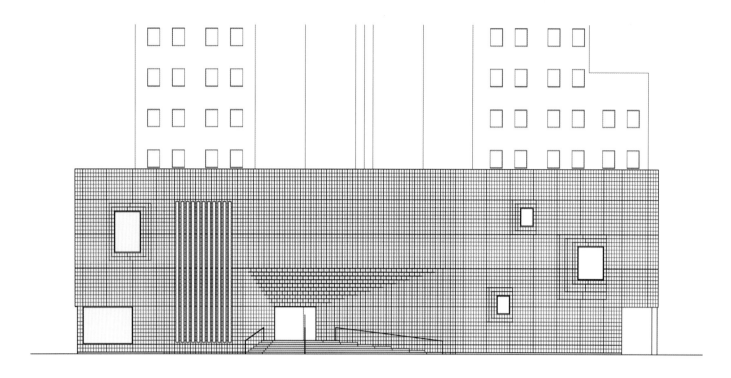

本项目位于上海常德路和南京西路交汇口一个大型高端商业区内，是宝姿 1961 的全新旗舰店的外墙设计。外墙代表的是宝姿 1961 品牌的未来愿景，是品牌起源和发展的结合。设计灵感来自漂浮在海洋上的冰山；建筑呈现波形外观，具有膨胀感、收缩感，似乎是被环境自然塑造出来的。外墙的造型展现出设计实验的种种可能，展现出形式、材料和技术的转换，同时表达对传统和现代的理解。建筑的宏伟结构引人注目，矛盾的本质则与环境积极互动。

外墙由两种玻璃材料组成：标准的 300 毫米 x300 毫米玻璃和定制的 300 毫米 x300 毫米 x300 毫米转角玻璃。与大部分玻璃砌体结构不同，外墙并不局限于一个垂直平面。两种玻璃材料的组合使用形成一个具有雕刻感的三维外墙，展示悬臂结构。加入创新的结构工程，打造新的连接系统，实现精心装饰的阶梯阳台，以自然的角度朝向步行人群，从各个角度都可以看见四扇凸肚窗。

丝光玻璃材料和喷砂处理的不锈钢材料与混乱的城市形成了鲜明的对比。白天的时候，店面为周围环境带来一丝静谧感，巧妙地反射着日光。夜幕降临时，景色变得冰冷干脆，嵌入的 LED 灯将外墙点亮。外墙上不同的几何形状和变化的视角表达出上海这个城市和人口多变的特征。

外墙承包商：J. Gartner（香港）有限公司　　面积：1145 平方米　　施工时间：2014 年 -2015 年　　摄影：苏圣亮

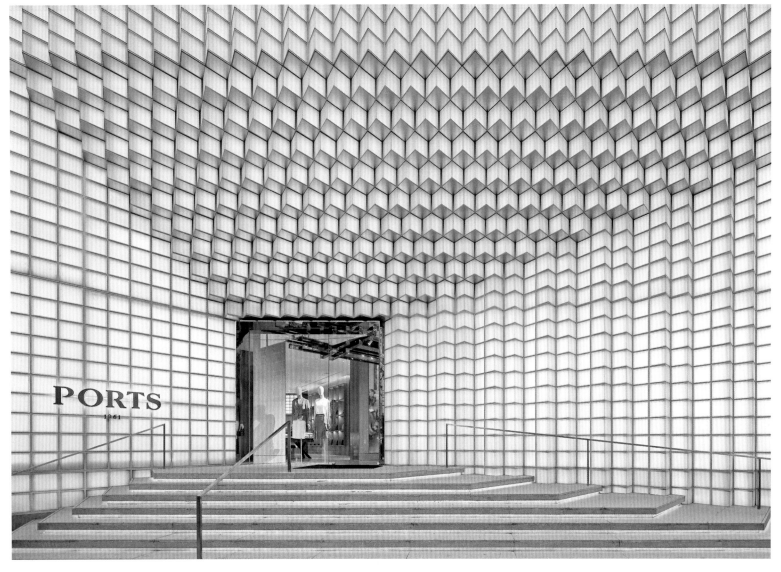

COLKITT&CO

PUMA FACTORY OUTLET CENTER (FOC) ATLANTIC

Atlantic City, NJ, USA

大西洋城彪马工厂直销中心（FOC）

内森·李·科尔基特（科尔基特设计公司）／美国，新泽西州，大西洋城

TColkitt&Co's model is revolutionizing PUMA's global outlet program with mobile fixture kits that allow for more flexible and sustainable retail design. Between interchangeable stock parts made of 100 percent recycled materials, to a 35 percent more energy efficient design, the PUMA factory outlet center (FOC) model is setting a precedent for LEED Platinum retail design.

The FOC model by Colkitt&Co is a very universal design, but also thoughtful based on PUMA's considerable research of their own carbon footprint. From a construction, design, sustainability, and investment perspective, all elements were purposely selected to improve shopability for the end customer and foster a better lifecycle of materials. This 'design by subtraction' concept is based on introducing a mobile fixturization kit of interchangeable stock parts that yield 30 percent lower energy consumption and fewer materials from the lack of partition walls, enabling the entire store to be reconfigured without the use of permanent construction or installations. This prototype for PUMA also provides extreme flexibility for seasonal and inventory needs, as the store can be packed up quickly and relocated to another site with minimal deconstruction and waste. In regards to some of the materials used, all wood is renewable, all metals are recycled (such as hot rolled mild steel), and both Plyboo and Wheatboard are FSC-certified. LED lighting is placed at a 45 degree diagonal pattern, significantly reducing the lighting quantity and power load, making this design 35 percent more energy efficient than the California Energy Code and American National Standard.

科尔基特公司的目标是使用可移动的灵活和可持续零售设计重新打造彪马的全球工厂直销店。百分百回收材料制成的可替换结构，能效提升35％的设计——这间彪马工厂直营店（FOC）设定的是LEED白金级零售设计的一个先例。

科尔基特设计公司打造的这间彪马工厂直营店具有较高的通用性，也充分考虑到了彪马对品牌自身产生了碳足迹进行的大量研究。从建筑、设计、可持续性和投资的角度分析，所有的元素选择都有意地关注终端顾客的消费体验，更好地实现材料的循环利用。这个"减法设计"概念的基础是引入可移动的可替换结构组合，使能耗降低30％，通过舍弃隔断墙减少材料消耗，使得整个店面的安排设计中没有永久性结构或装置。这一设计也为季节性和库存的需求提供了极大的灵活性，店面可以快速整理迁移到其他地点，将拆除工作和浪费降低到最小程度。至于使用到的一些材料，所有的木料都是可再生的，所有的金属均为回收金属（如热轧低碳钢），竹板材和麦草刨花板均有FSC认证。LED照明装置位于45度斜角图案，大大降低了照明装置的数量和功率，使得这个设计的能效比加州节能标准和美国国家节能标准提高了35％。

Designer: Nathan Lee Colkitt Area: 5,332 SF Photographer: Garrett Rowland

设计师：内森·李·科尔基特 面积：5,332平方英尺 摄影：加勒特·罗兰德

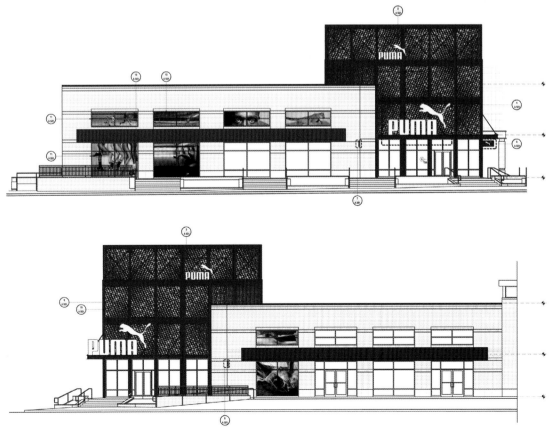

HANGZHOU GUAN INTERIOR DESIGN

SAN SAY WOMEN'S WEAR

Hangzhou, China

San Say（桑索）品牌女装

杭州观堂设计 / 中国，杭州

The definition of San Say is French styled clothing for women, joyful and pleasant. The client wishes to improve interest in the shopfront design for a new generation of branding image.

The designer uses fun and interest throughout the entire shopfront in order to improve customer experience. Through the large glass windows, one can spot a number of inverted furniture and decoration, including floor lamps, sofas, tea tables, and chairs; all arranged against common sense. Some are rested against the wall, some are against the ceiling; decorative lamps on the wall no longer focuses on the expression of texture, but decorated with two lines of light paint, reminding of the shadows cast by the lamp; on approaching the displayed garments, the customer will find her own face reflected in the bronze mirror in front of her. That is her new image with the new clothing 'on', just like a princess in the fairy tale. The façade employs an impressive colour scheme of turquoise. Varying shades, together with simple outlines indicate a vivid Neo-classical elegance.

Designer: Jian Zhang Property: commercial, branded clothing Completion Date: June 2014 Area: 130 sqm Photography: Yujie Liu Description: Tang Tang Materials: wood texture, linear outline

San Say（桑索）定位于法式女装，品牌充满愉悦与欢快，客户在新一代形象设计中希望能增加店铺的趣味性。

设计师接手本案例后，为增加客户体验感，将趣味性贯穿整个店铺。透过宽大的玻璃窗可以看到，在店铺中有很多倒置的家具和饰品，例如落地灯、沙发、茶几、椅子，统统打破常规摆放，进行空间倒置，有的上墙、有的爬上天花板；墙面的装饰壁灯下，不再是材质做表述，而是刷上淡淡的两色漆，仿佛壁灯投下的光影；走近展示的衣服，抬头，猛然发现有枚铜镜里印衬的正是自己的脸庞，配上店里的新衣，仿佛童话里的公主。外部门面在色彩选择上采用了特殊的浅浅的孔雀蓝，清新、淡雅，配合店铺简洁的欧式线条，新古典主义风范呼之欲出。

设计者：张健　空间性质：商业空间，品牌服装　完工时间：2014年6月　面积：130平方米　摄影：刘宇杰　撰文：汤汤　主要材料：木纹，线条

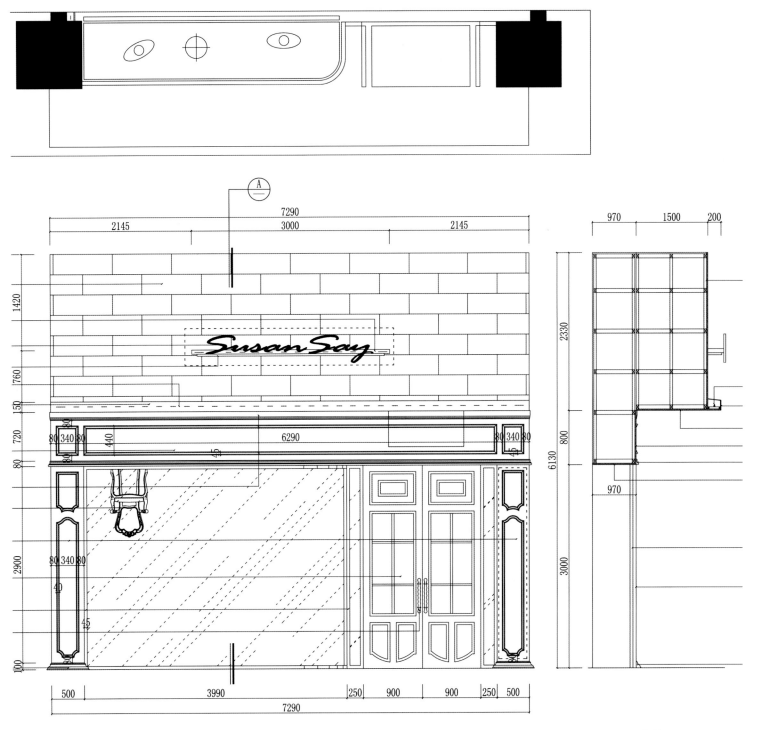

HANGZHOU GUAN INTERIOR DESIGN

SWEET BASIL WOMEN'S WEAR

Hangzhou, China

紫淑品牌女装

杭州观堂设计 / 中国，杭州

Sweet Basil, an exquisite herb originated from Europe, has a uniquely refreshing fragrance, which is warm and calm, pure and welcoming, last but not least, in a natural manner. It gives off a fragrance that people immerse themselves in. Sweet Basil wishes to help women explore the unique scent for themselves.

Smart, elegant with a tint of loveliness: this is what Sweet Basil wants to express in its shopfront. It uses a general naturally light colour scheme and the warmth of wooden tone employed in its furniture and fixtures to create a welcoming atmosphere. European styled elements can be found in the exterior, where the exposed texture of bare bricks display the impressive aesthetics of the brand.

Designer: Jian Zhang   Completion Date: January 2014   Area: 170 sqm   Materials: tile, wood, floral art   Photography: Yujie Liu   Description: Tang Tang

Sweet Basil，紫淑，源于欧洲的美妙香草，独有一种沁人心灵的香味，她的芬芳温暖而淡定、亲和而纯净、成熟自然不做作，低调的让人在她的香氛中沉浸。紫淑品牌想为女性发掘出属于自己的不可缺少的独特女人香。

小资、知性、不失纯真可爱，便是紫淑所追求的氛围，店铺整体处理为浅色调，干净雅致，配以木色的温馨，包括各类道具也采用木色+白色的结合，营造柔和温暖的气氛。外观运用了欧式风格元素，一部分裸露的砖墙纹理，展现出品牌的独特品味。

设计者：张健 完工时间：2014年1月 面积：170平方米 摄影：刘宇杰 撰文：汤汤 主要材料：花砖、木质、花艺

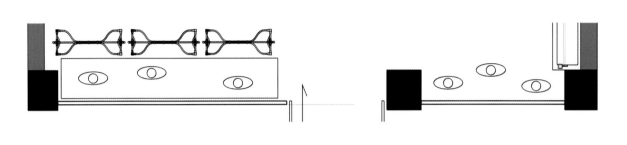
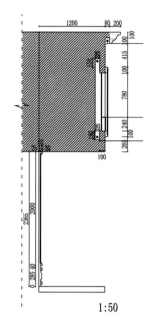
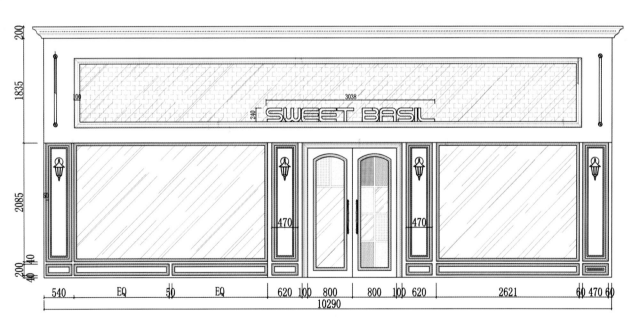
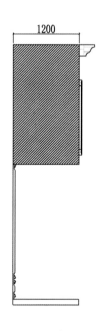

MVRDV

CRYSTAL HOUSES

Amsterdam, the Netherlands

水晶屋

MVRDV 事务所 / 荷兰，阿姆斯特丹

The entirely transparent façade of a high-end flagship store on Amsterdam's upmarket shopping street, PC Hooftstraat, uses glass bricks, glass windows frames and glass architraves in a way to evoke the vernacular of the area with the goal to maintain the character of the site. The 620 m^2 of retail and 220 m^2 of housing, were designed for investor Warenar. The design unites the ambition of Amsterdam to have large distinctive flagship stores without compromising the historical ensemble.

MVRDV's Crystal Houses began its existence with the request of Warenar to design a flagship store combining both Dutch heritage and international architecture on the PC Hooftstraat, Amsterdam's one and only luxury brand street that was previously primarily residential. MVRDV wanted to make a representation of the original buildings and found a solution through an extensive use of glass. The near full-glass façade mimics the original design, down to the layering of the bricks and the details of the window frames, but is stretched vertically to comply with updated zoning laws and to allow for an increase in interior space. Glass bricks stretch up the façade of Crystal Houses, eventually dissolving into a traditional terracotta brick façade for the apartments (as stipulated in the City's aesthetics rules), which appears to be floating above the shop floor.

Complation Date: 2016   Photographer: Daria Scagliola & Stijn Brakkee

Original Façade
原有外墙

Existing Situation
保留结构

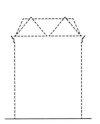
Houses Replaced with Larger Volumes
扩大建筑体量

Old Pleiter Façade Rebuilt in Glass
旧外墙进行玻璃改造

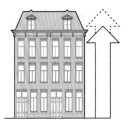
Facade Stretched to Fit New Volume
外墙延展适应新体量

Glass to Terra Cotta Brick Gradient
玻璃到红土砖渐变效果

坐落在阿姆斯特丹的高档购物街 PC Hooftstraat，这座高端旗舰店为了维护此基地特征，采用玻璃砖、玻璃窗框和玻璃框缘而成的完全透明外墙立面，唤起人们对块区域的乡土之情。本项目是受开发商 Warenar 委托而设计的 620 平方米零售空间和 220 平方米住宅空间。设计方案结合阿姆斯特丹的野心，在兼顾旗舰店精彩设计的前提下，又不折损其历史元素。

MVRDV 的水晶屋服装店设计从开发商 Warenar 的委托开始。开发商希望在阿姆斯特丹这条原为住宅而现今唯一的精品街 PC Hooftstraat 上，打造一间结合荷兰传统与国际化建筑风格的品牌旗舰店。设计师希望做出体现原有建筑风貌的店面设计，并通过大量使用玻璃建材找到了解决方法。这座几乎全玻璃的外墙设计下至砖石的分层和窗框的细节都模拟了原来的设计，但垂直方向进行了延伸以符合最新的分区法规，同时增加内部空间。水晶屋的外墙通过玻璃材料得到了延伸，最终与公寓的传统陶土砖墙相融（符合城市的美学规范），使得公寓看起来像是飘浮在店面上方。

项目时间：2016 年　摄影：Daria Scagliola & Stijn Brakkee

MASAFUMI TASHIRO DESIGN ROOM

ANSHINDO PARIS

Rue de la Paix 8 Paris, France

ANSHINDO 品牌巴黎店

田代雅文设计工作室 / 法国，巴黎和平街 8 号

The Jewellery shop ANSHINDO is located in Paris Rue de La Paix 8. The major renovation of a grand old molding decoration building facade has resulted in intact Paris' traditional and transparence shop front.

The idea of creating a new shop front for the old shop Anshindo, the designer used the transparent forming glass moldings glued on the glass surface that gave mix with new and traditional form.

While people walking around in front of the store, it brings the glow of diffuse reflection and refraction to the glass facade, it resulted beautiful natural and organic scenario with its occasion.

Designer : Mr. Masafumi Tashiro   Area: 70m²   Photographer : Mr. Yosuke Kojima

Anshindo 珠宝店位于巴黎和平街 8 号。这项针对巨大旧式装饰建筑外墙的主要改造工程保留了巴黎传统风格，同时打造出透明的店面外观。

为 Anshindo 打造新店面的设计中，设计师选用透明玻璃造型粘在玻璃表面上，构成新旧交替的形式感。

行人在店面门前行走时，玻璃外墙上会呈现出反射和折射形成的光泽，打造出美好而天然的购物环境。

设计师：田代雅文　面积：70 平方米　摄影：小岛阳介

KOIS ASSOCIATED ARCHITECTS

ILEANA MAKRI STORE

Athens, Greece

伊利安娜·马克里店面

科伊斯联合建筑师事务所 / 希腊，雅典

On the exterior the shop window opening was subdivided into three main areas. Two stone framed windows protruding form the façade were created and the entrance door was encased between them. The first framed window is the exhibiting window that holds the new collections. This opening frames a metal-glass tree with hanging exhibition cases and the stores stone clad staircase leading to the jeweller's workshop. The passer's gaze is allowed to penetrate the interior and capture glances of the movement within and observe the people ascending and descending the levels of the store. The second stone rimmed opening frames an art installation. The entrance door that is encased between the two frames is made of black coloured oak wood. The designers wanted the door to be an element of the façade and not just an opening. Its height and robust construction create a monolithic appearance which combined with the dark colour of the wood allude to the door to an enchanted forest. With the asymmetric subdivision and stereotomic properties of the synthesis the designer team gave the façade a rhythm and coherent composure. The designers wanted to reveal not everything at once but hide aspects of the shop in order create ambiguity and depth, curiosity that would lead to a stasis in a street of constant movement.

Within the art installation window smoke is released through hidden outlets and creates a controlled cloud within the window. The designers wanted to explore a mode of fluctuating transparency by the use of slow released smoke and the formation of a nebula that would act as a veil. A constantly changing veil, which regulates the views from the exterior to the interior and the opposite, creates a sense of mystery and ethereal enclosure.

Principle Architect: Stelios Kois Project Manager: Antriana Voutsina Design Team: Nikos Patsiaouras, Marielina Stavrou, Konstantinos Karanasos, Alexandros Economou Photography: Giorgos Sfakianakis

店铺外墙平面上的窗口分成三个主要区域。外墙上突出两扇石质框架的窗口，入口正门位于二者之间。第一扇窗户是展示最新系列商品的橱窗。窗口框架内是一棵金属和玻璃组成的树，悬挂着展示商品，店内的石阶楼梯通往珠宝商的工作间。过往行人的目光可以到达室内空间，捕捉到内部的运动，观察到人们在店内上上下下。第二个石沿窗口内是一座艺术装置。入口大门位于两个黑色橡木窗口中间。设计师希望这扇门成为外墙的一项重要元素而不只是一个普通的开口。门的高度和结实的结构形成了统一的外观，结合木质的深色，让人联想起充满魔法的森林。设计师利用不对称的细分设计和立体切割特质，为外墙赋予节奏和连贯的沉静之感。设计师希望有所保留地展示店面元素，创造出模糊感和深度，以及吸引行人前来探索的好奇感。

艺术装置橱窗内的隐藏通道释放出烟雾，在窗口处形成一片可控的云雾。设计师希望通过使用缓慢释放的烟雾和类似面纱的星云结构探索波动的透明模式。不断变化的面纱使得由外到内以及对面的观察视线变得规律，构成神秘感和飘渺的店面印象。

主要建筑师：斯泰利奥斯·科伊斯　项目经理：安特里亚纳·福特西娜　设计团队：尼科斯·帕西奥乌拉斯，马里埃丽娜·斯塔夫鲁，康斯坦丁诺斯·卡兰纳索斯，亚历山德罗斯·伊科诺穆　摄影：乔治斯·法基纳基斯

CHRISTIAN LAHOUDE STUDIO

TIFFANY & CO – CHONGQING, CHINA

Chongqing, China

蒂芙尼中国重庆店

克里斯蒂安·拉胡德工作室 / 中国，重庆

This project is a complete renovation for Tiffany's jewellery store located in Chongqing, China. A full architectural intervention included a storefront design and interiors creating an environment connecting the brand's rich history and identity with today's customers in Asia.

The goal was to combine nature, which has always been a constant design theme for Tiffany, with modern architecture to create an uplifting experience within the three-story space. The illuminated site specific sculpture of falling leaves made of Tiffany blue and iridescent metals by Japanese artist Nami Sawada is viewable from the outside and spans the full length of the entrance, introducing the free-flowing state of the design before one even enters the store.

The classic elements of Tiffany & Co. flagship store on the fifth avenue are employed in this project while its typical symbolic patterns are found on the decorations: the façade has a stereoscopically engraved effect; where the pearly gloss is reflected on the metal exterior, resembling the white ribbon on the famous Tiffany Blue Box®; the frame of the fine steel store entrance is decorated with grain pattern representing nature.

本案是为位于中国重庆的蒂芙尼珠宝店进行的重新翻新项目。完整的项目包含了店面和室内设计，打造将品牌丰厚历史和品牌形象与当今的亚洲消费者连接在一起的购物环境。设计目标是结合蒂芙尼一直沿用的设计主题——自然元素，通过现代建筑在三层空间内营造出振奋人心的体验。采用蒂芙尼蓝和日本艺术家泽田奈美制作的闪光金属特别打造的落叶雕塑从店面外部就可以观察得到，跨越整个入口宽度，顾客在进店之前就可以充分感受到自由流动的设计精神。

该项目运用了品牌位于第五大道旗舰店的经典元素，同时延续极具代表性的符号图案装饰：外立面呈现立体雕刻效果；建筑使用的金属表面饰以珍珠色泽，宛若包裹著名的蒂芙尼（Tiffany & Co.）蓝色礼盒（Tiffany Blue Box®）的白色缎带；精钢材质的专卖店大门，外框上装饰有象征自然的麦穗图案。

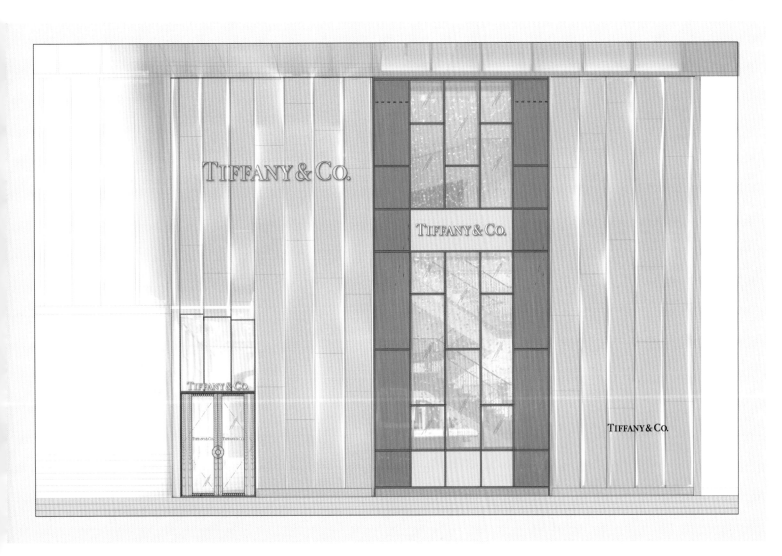

OSCAR VIDAL STUDIO

BIG CHANES SALAOS – TORREVIEJA

London, UK

大查内斯·萨拉斯——托雷维耶哈

奥斯卡·维达尔工作室 / 英国，伦敦

The new store is a step further in the expansion of the company Chanes Salaos. This is the third store and the biggest, with it they wanted to be a reference in their market in their area. BIG CHANES SALAOS is located in Ozone Leisure Center and was characterized by a facade to three big faces with many windows. For it Oscar Vidal Studio designed a new logo to communicate the spirit of the business mixing iconic symbols of toys such as the teddy bear, circus and soldiers. As key solutions they chose to create a series of iconic elements to populate the façade concerning the world of toys as soldiers and stuffed animals and circus like tents and tarps red and white bands. The main active was the splendid façade, very long 5 meters tall and with 3 faces located on the corner of the mall. So it was treated as a scene with the teddy bear logo in the middle protected with a soldiers guard. The windows were treated with striped curtains with circus like aspect enclosing a typical toy scene as if it was a theatre cage. The design team created several scenes with different topics: balloon, crane, party, podium, and boxes.

Area: 225m² Completion Date: June 2014

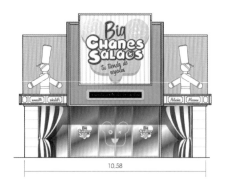

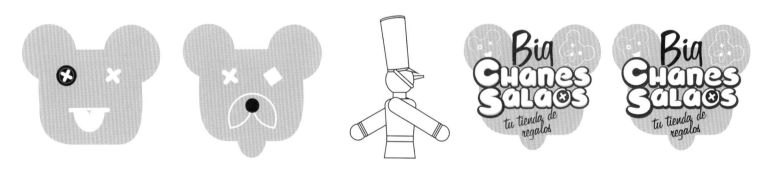

这间新店是查内斯·萨拉斯公司扩张过程中的重要一步，它是公司的第三家店，也是最大的一间。品牌希望借此店铺打造同类市场里的典范。大查内斯·萨拉斯位于 Ozone 休闲中心，店面外墙上的窗户设计让人过目不忘。设计团队为其打造了全新标识，结合泰迪熊、马戏团和士兵等经典玩具符号传达商业精神。作为核心设计方案，设计团队选择打造一系列标志性元素丰富外墙设计，采用士兵和填充动物等玩具世界元素和帐篷、红色油布和白衣乐队等马戏团元素。作为店面设计中的主要元素，极长的 5 米高墙壁和商场角落的三面场景组成华丽的外墙。泰迪熊的标识在中间，被一个士兵守护，构成了一处场景。透过窗户上的条纹窗帘，典型的玩具场景若隐若现，似乎里面是一个剧场。设计团队打造了不同主题的多个场景：气球、起重机、聚会、领奖台和盒子。

面积：225 平方米　建成时间：2014 年 6 月

CURIOSITY

MAISONMOYNAT, LONDON

London, UK

伦敦摩纳大厦
CURIOSITY 公司 / 英国，伦敦

In one of the most charming neighbourhoods of London, at 112 Mount Street facing the Connaught Hotel, the London store is the first Moynat shop to open outside of Paris, after the original shop in Paris Saint Honore.

With each of Moynat's new stores the designer explores the different facets of the brand's personality, though always focusing on the ultimate refinement expressed with a minimum of means: continuity, genuinity, and singularity.

The domed ceiling and the arched windows of the store echo the signature curves on Moynat trunks and handbags. The pattern of the ceiling is inspired by the Art Deco period, the golden age of Moynat, when the artist Henri Rapin was the creative director of the House. The unique lighting creates a sense of refinement with a limited almost minimalist palette of materials: leather, suede, wood and strokes of metals, all inspired by the products themselves.

Completion Date: March 2014   Area: 160m²   Photographer: Melvyn Vincent

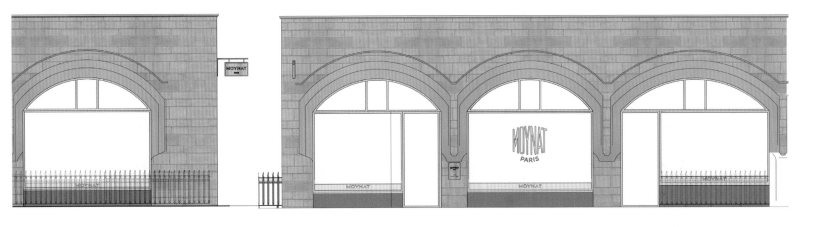

摩纳在伦敦的第一家店坐落在伦敦最迷人的街区之一，位于蒙特街112号，对面是康诺特酒店。这是摩纳品牌在巴黎圣奥诺雷的原始店以外开设的首个店面。

设计师在摩纳的每个新店铺中探索品牌个性的不同方面，即便一直关注简约设计手法表达的终极精细：连续性、真实性和独特性。

店内的穹顶设计和拱形窗户呼应摩纳衣箱和手袋标志性的曲线设计。天花板上的图案灵感来自摩纳品牌的黄金时期——装饰艺术时期，那时由艺术家亨利·列宾担任品牌的创意总监。独特的照明设计打造出材料使用中有界限的，近乎极简主义之感：皮革、麂皮、木材和局部金属。所有的设计灵感都来自产品本身。

建成时间：2014年3月　面积：160平方米　摄影：梅尔文·文森特

TORII DESIGN OFFICE

DOUBLE SAKAE

Nagoya, Japan

DOUBLE 旗舰店

鸟居佳则设计事务所 / 日本，名古屋

This store is located in the center of the city. Designers were asked to evoke a clean image for the shopfront, which must attract the attention of passers-by. Then the idea of dying all over the building pure white, the exterior and the interior is preferred. According to Japanese saying, the colour of white is way more than 100 colours. The existing building is a financial office building of two-storey wood structure built in the 1940s, and the structure is newly enhanced with the steel frame. The client wants to bring life back into the old building and make it a select shop.

Exterior elevations are wrapped with white iron panels, and people will never recall the wood structure. There is a large opening of transparent glass wall in the first floor, and the exterior white wall in the second floor is stamped with the logo of DOUBLE with three-dimensional effects. The RGB LED lighting is adopted, and the color of lighting varies from white to blue. Linear illumination is installed along the margin of the building and the store will stand out at night.

Designer: Yoshinori Torii   Completion Date: May 2014   Area: 264.46m²   Photographer: Nacasa & Partners inc.

店面位于城市中心地带。设计师受邀打造一个洁净的店面形象，必须有效吸引路人的注意力。设计师选择将室内外粉刷成纯白色，因为日本人认为，白色的美好胜过百种颜色。项目中原有建筑是一栋金融办公楼，是一处建于20世纪40年代的两层木质结构，近年进行了钢框架加固。客户希望通过这次改造工程让原有建筑获得重生，并将其打造成一个精选商店。

店面外墙被白色铁板包围，人们看不到原来的木质结构。一楼宽大的透明玻璃墙以及二楼的白色外墙上都贴有三维立体效果的店铺商标。三色LED照明装置可以发出从白到蓝的多色灯光。沿建筑和店面边缘的直线照明装置会在夜间点亮。

设计师：鸟居佳则　建成时间：2014年5月　面积：264.46平方米　摄影：Nacasa & Partners公司

SUITE ARQUITETOS

BOTTI STORE

Jardim, Sao Paulo, Brazil

博蒂鞋店
Suite 建筑事务所 / 巴西，圣保罗，雅尔丁

For Botti store renovation Suite Architects used the principles of upcycle. The main idea of the facade is the reversion of the shoes' functions and concepts.

The facade of the old house was covered in wooden planks, forming a 'big box', distinguishing from the inside atmosphere where everything is white and very conceptual.

Completed Date: May 2014   Photographer: Ricardo Bassetti

博蒂鞋店翻新项目中，Suite 建筑事务所采用了升级改造的设计原则。外墙的主要概念围绕鞋履功能和概念的逆转而展开。

旧店面的外墙由木板覆盖，形成一个"大盒子"的外观，在所处的全白色抽象环境中脱颖而出。

建成时间：2014 年 5 月   摄影：里卡多·巴塞蒂

CHRISTIAN LAHOUDE STUDIO

JIMMY CHOO CHENGDU, DACI TEMPLE

Chengdu, China

成都大慈寺周仰杰旗舰店

克里斯蒂安·拉胡德设计工作室 / 中国、成都

Jimmy Choo's 140-square-meter flagship in Chengdu, China is a two-level, dual gender store that introduces itself with an 'open' façade, the upper part of which reveals the activity in the store. The lower section of the façade is more complex, with rounded edge cartouche framings that marry retro and modern design through lighting installed within the frames. Marble and metal layering is deconstructed to give way to the light from inside the store.

Draped in white Carrera marble, the welcoming feature stairway continues the theme of transparency as it wraps around the showcased product. The decor is luxurious but warm, featuring gold mesh panels on walls, a chandelier at the entrance with burnished gold hoops and raw crystals, marble floors, rich grey carpeting, and grey-pink walls. Delicate glass globes are suspended over shelves and mirrors that tilt are positioned above shoe displays.

Completion Date: January 2015   Area: 139m²   Photographer: Eric Gregory Powell

周仰杰成都旗舰店占地140平方米，是一个两层的店面，店内男鞋女鞋均有销售。"开放式"店面吸引消费者进店探索，上层空间是店内活动的主要场所。店面外墙的下层部分设计更为复杂，照明装置配合线条丰满的漩涡装饰框架结合了复古和现代设计元素。大理石和金属的层次呈现解构概念，让位于店铺内部的光线。

入口的特色楼梯采用白色卡雷拉大理石饰面，延续围绕展示商品的透明设计主题。室内装饰呈现奢华而温馨的风格，墙壁使用金色网状面板，入口处的吊灯上是光亮的金色吊钩和半加工水晶，另外还有大理石地面、厚实的灰色地毯和粉灰色墙壁。精致的玻璃球体悬挂在货架和镜子上方，照耀精美的鞋履。

建成时间：2015年1月　面积：139平方米　摄影：埃里克·格雷戈里·鲍威尔

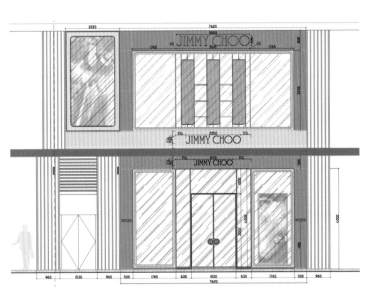

FMO ARCHITECTURE

THE LISBON WALKER FLAGSHIP STORE

Lisbon, Portugal

里斯本沃克旗舰店

FMO 建筑工作室 / 葡萄牙，里斯本

The Lisbon Walker, a totally new premium concept brand, centrally located in the downtown of Lisbon, is where the very best of Portugal comes together, mainly in the art of men shoes and wine, besides belts, accessories and concept items, entirely designed and produced in Portugal.

The goal was to create a space where products could be shown in a clear, objective and true way. Exhibiting shoes and wine bottles side by side in the simplest way possible, by making them float on the walls and in the storefront, was the architectural concept. Symmetry, simplicity and the truth of materials were the main ideas for both space and furniture designs.

The storefront presents us a clear approach to tradition and modernity, joining architecture and the product concept together. On the left side of the entrance, the show window captivates people's attention already with the product concept: shoes and wine. Nevertheless, the design of the showcase enables passers-by to look deeply into the store and pulls them inside.

Designer: Filipe Melo Oliveira Completion Date: May 2014 Area: 59m² Photographer: © João Morgado – Fotografia de Arquitectura

里斯本沃克是一家全新的高端概念品牌，店面位于里斯本市中心的核心位置，这里聚集了葡萄牙最优质的品牌和店铺，主要包括男鞋和红酒、腰带、饰品和概念产品，全程在葡萄牙设计生产。

本案的设计目标是打造一个清晰、直观、真实的商品展示空间。设计师选择了最简洁的方式同时展示鞋履和红酒，采用飘浮在墙壁和店面上的建筑理念。对称性、简约性和真材实料是空间和家具设计的主要概念。

店面清楚地向人们呈现出传统和现代的设计理念，将建筑和产品概念联系在一起。入口左侧的展示橱窗呈现的产品理念能够高效地引起行人的注意：鞋履和红酒。在此基础上，展示橱窗的设计吸引人们更为深入地探索店铺。

设计师：菲利浦·梅洛·奥利维 建成时间：2014年5月 面积：59平方米 摄影：© 若昂·莫尔加多——Fotografia de Arquitectura 公司

ARNAU VERGÉS TEJERO

L'ÒPTICA CLARÀ D'OLOT

Catalonia, Spain

"l'ptica Clar à d'Olot" 眼镜店改造工程

阿尔瑙·维杰斯·德杰罗 / 西班牙、加泰罗尼亚

The designers set off from the existing store, with generous windows framed with natural wood. They do not intend to start from the scratch: they believe in the value of a well performed job and adopt the existing frames as if they were their own. Based on the frames made of 7×7 cm wood strips, the designers make up a 3D structure as a kind of scaffolding that joins the strips laterally and duplicates the mesh of the facade inside the shop. This way they generate a structure with a porch, or an atrium, which embraces the inner door in the center, the display windows at each side, and structural pillars or elements of installations in the boxes formed between the empty spaces.

Photographer: Marc Torra Ferrer

设计师从原有店面的大窗结构入手，在7厘米×7厘米的木条围成的窗框上打造出三维立体结构，横向连接木条，同时呼应店内的网状结构。如此形成的玄关或中庭结构便将店面中央的内部大门、每侧的展示橱窗和结构立柱联系在一起。

摄影：马克·多拉·费雷尔

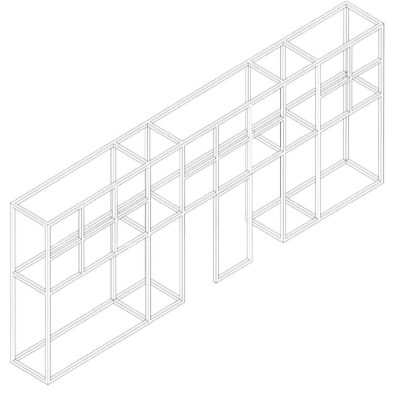

MASQUESPACIO

KESSALAO

Bonn, Germany

凯莎拉奥餐厅

Masquespacio 设计公司 / 德国，波恩

Spanish creative consultancy Masquespacio presents their last project realized in the city of Bonn, Germany. The project consists in the brand image and interior design for Kessalao, a new take away establishment of Mediterranean food in the city of Beethoven.

In so far as the shopfront and interior design it's presented by a space that symbolizes the freshness of the brands' name through a range of most popular colours for Germans. Red is without a doubt the main colour, while the marine blue and yellow remind us of the Mediterranean Sea. Purple on the other hand adds a strong touch to the whole together with the principal red colour. Materials like wood coming from the birch veneer used for the walls and pine for the furniture, where chosen to offer a natural look to the space. Moreover through different decorative elements made of raffia as for the seats and pots a Mediterranean touch is added repeatedly. Masquespacio with this project starts his first international adventure, added to a project of shopfront and interior design actually in development in Oslo, besides several projects in progress of branding and consulting in Spain, showing its ability to offer creative solutions for any business model, adapted to its target audience.

Creative Director: Ana Milena Hernández Palacios   Architect Junior: Virgínia Hinarejos   Graphic Designer Junior: Ana Diaz   Area: 40m²   Photographer: David Rodríguez y Carlos Huecas

西班牙创意咨询公司 Masquespacio 的最新项目位于德国波恩。该项目包含品牌形象设计和室内设计，委托方是贝多芬市的新创立的地中海食品外卖品牌凯莎拉奥。

店面和室内设计中使用了一系列德国人最喜爱的配色表现象征品牌新鲜度的空间概念。红色无疑是设计方案中的主色，海军蓝和黄色让人联想到地中海。紫色则在主要的红色基础上为整体增添了不同的韵味。设计师使用了桦木墙面、松木家具，增添自然气息。酒椰叶纤维制成的多种装饰元素，座椅和坛坛罐罐重复体现了地中海风情。Masquespacio 设计公司通过这个项目开启了国际化的旅程，通过此项目以及奥斯陆正在建设中的一个店面和室内设计，几个正在进行中的西班牙品牌设计和咨询项目展示出公司为任何商业模式提供针对目标受众的创意设计的实力。

创意总监：安娜·米莱娜·埃尔南德斯·帕拉西奥斯　初级建筑师：弗吉尼亚·海纳拉赫斯　初级平面设计师：安娜·迪亚兹　初级创意设计：卡罗琳娜·米科　面积：40平方米　摄影：大卫·罗德里格斯，卡洛斯·韦卡斯

NERI&HU DESIGN AND RESEARCH OFFICE

RACHEL'S BURGER

Shanghai, China

Rachel's 汉堡餐厅

Neri&Hu 设计事务所 / 中国，上海

Drawing inspiration from the American drive up burger joints of the 1950's, the restaurant was envisioned as a porous space where the boundaries between inside and outside are blurred.

The exterior walls can be fully opened to further extend the perceived and actual boundaries of the restaurant. When fully closed, however, a clever combination of clear and textured glass along with mirror are used to great effect to visually extend the boundaries of the space while bringing light, views and streetscape deep into the interior.

The dematerialization of the walls is further emphasized by the dominance of the horizontal planes.

The roof structure seemingly floats above the space while the floor rises and falls to support the custom eating and seating surfaces, integrating communal tables with pivoting benches accommodating for individual or group dining flexibility.

Area: 93m^2 Materials: Custom steel and glass doors (steel, clear glass, textured glass, mirror), hand painted concrete tiles

受到二十世纪五十年代美国街边连锁汉堡店的启发，Rachel's 汉堡餐厅采用了通透的设计理念，模糊了餐厅室内外之间的界限。

完全开放式的外墙进一步延伸了餐厅的边界。全部闭合后，光面玻璃、纹理玻璃与镜子的搭配使用不仅能够在视觉上留有更多空间，也为室内带来了更加充足的自然光和全景式的室外视野。

餐厅没有使用实体墙，轻盈的结构通过狭长的平面得到进一步加强。

从视觉的角度，屋顶结构似乎悬浮在空间之上，地面和就餐吧台采用统一的几何图案，有如地面动态地上下起伏。长桌和旋转吧台椅可以灵活满足不同就餐人数的需求。

面积：93 平方米　材料：定制玻璃钢门（钢、透明玻璃、纹理玻璃、镜子）、手绘混凝土砖

THE SWIMMING POOL STUDIO

HEIYU NONG FISH THEMED RESTAURANT

Nanjing, China

嘿鱼弄特色鱼餐厅

上海三也室内设计 / 中国，南京

'Heiyu Nong' Fish Themed Restaurant is a young light meal restaurant advocating natural popular dining philosophy. The design inspiration is drawn from the main ingredient—fish. The designers take an abstract approach to the full image of fish scale and reinterpret it in metal finish to create a scale-like impression. The indoor dining area is a combination of a lot of recycled timber, metal panels and textured concrete walling, which is also reflected in the entrance elevation. Wooden background of the restaurant logo adds to the natural ambience of the entrance. Units of scale style decorative panels in 60cm×60cm form the elevation. Repetition of the same elements creates a striking visual impact. The indented form of entrance functions as a leading guide to the traffic flow. The old timber here reminds of the materials and textures found inside the restaurant; the use of large area of glass makes the whole entrance more permeable. All the design details work together to form a vivid restaurant image of 'fish'.

Dining is not just eating, but a complex journey companied by the aesthetics and comfort provided by space, tableware, menu, lighting, and furniture. Therefore a successful shopfront design is crucial to creating an impressive branding identity.

Designer: Linjie Li Completion Date: April 2015 Area: 200m² Photography: Wenjie Hu

"嘿鱼弄"特色鱼餐厅是一家定位为自然时尚的年轻饮食主义餐厅。设计灵感来源于餐厅的主要食材——鱼。设计师将鱼鳞匀致饱满的形象抽象化，并用金属材质重新演绎出来，塑造了具有鳞片肌理的入口印象。餐厅的室内设计大量运用回收老木、金属板和肌理水泥墙面的结合。这一点在入口立面设计上也得到了体现。餐厅Logo的老木背景使整个入口形象变得更为古朴自然。立面大部以60cm×60cm大小的鱼鳞片造型装饰拼接而成。相同元素的重复使用创造了具有冲击力的视觉形象。内凹的入口形态对人流产生一定的导向性。而老木的使用与店内的材质相呼应；大玻璃的采用使得整个入口更为通透。所有设计细节的融合塑造了个性鲜明的"鱼"餐厅形象。

餐饮，不仅仅是吃饭，用餐者对空间、餐具、菜单、灯光、家具的美感、舒适度都有很高的要求。所以好的店面设计对品牌打造会有事半功倍的效果。

设计师：李麟杰 完工日期：2015.4 面积：200平方米 摄影师：胡文杰

081 • 餐厅

IGNACIO CADENA

HUESO MEXICAN RESTAURANT

Jalisco, Mexico

墨西哥韦索新概念餐厅

伊格纳西奥·卡德纳 / 墨西哥，哈利斯科州

Luis Barragan's Foundation as well as Diaz Morales's House-Studio in beautiful Lafayette Design District located in Guadalajara, Jalisco, Mexico serve as the perfect backdrop for a recovered 1940's modern architecture 240 sq. ft. building that would become Alfonso Cadena's new concept restaurant 'Hueso' (Bone).

The design approach begins with creating a double skin, in the exterior, a clean artisanal handmade ceramic tile covering with a graphic approach protects the inside skin layer which becomes more organic and full of texture. Inspired in a Darwinian vision, the inside skin layer represents the unique decoration of interior space which almost every vertical square inch of the interior covered with over 10,000 collected bones from animals and plants mounted on wooden layers, mixed with objects and cooking tools and intervened by urban visual artists.

Completion Date: 2014 Culinary Concept: Alfonso Cadena Concept & Art Direction: Ignacio Cadena Architect of Record: Javier Monteón Graphic Design: Rocío Serna Photography: Jaime Navarro

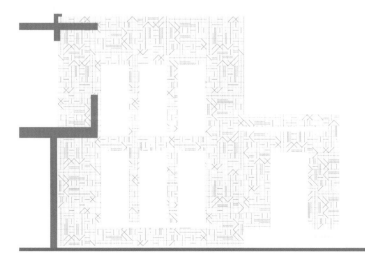

在墨西哥哈利斯科州的瓜达拉哈拉，美丽的拉斐特设计区有路易斯·巴拉干的作品和迪亚兹·莫拉莱斯的住宅／工作室，是一间始建于20世纪40年代的重建建筑的完美背景。这块240平方英尺的空间被选为阿方索·卡德纳的新概念餐厅"韦索（骨头）"的所在地。

设计方案以双层外墙为基础，外墙上洁净的手工瓷砖覆面平面设计保护内部墙面，同时组成一个更为有机，具有质感的结构。受达尔文主义启发，内层墙面代表室内空间的独特装饰。在几乎每个平方英寸的垂直室内空间都同样覆盖着超过10,000个收集而来的动物骨骼和植物结构。这些元素安装在木板上，其间混合着城市视觉艺术家的妙手处理过的物件和烹饪工具。

建成时间：2014年　美食概念：阿方索·卡德纳　设计理念与艺术指导：伊格纳西奥·卡德纳　建筑设计：哈维尔·蒙特昂　平面设计：罗西欧·塞尔纳　摄影：杰米·纳瓦罗

MINAS KOSMIDIS (ARCHITECTURE IN CONCEPT)

BIRIBILDU FAST CASUAL DINING

Athens, Greece

Biribildu 休闲快餐店

米纳斯·科斯米迪斯（Architecture in concept）／希腊，雅典

Biribildu in the language of the Basks means....

It is about a fast food restaurant with tasted reports in the Greek kitchen and designing reports in the wheel and in everything that rounds around itself. It rounds like 'gyros' (Greek specialize) which is the main product in the menu of the fast food restaurant. There are reports in the magical world of the circus and the carousels with whom a scene in a place of 70m (square) with view to a coast area of Athens. The red and the yellow colours, colours-symbols for the circus appear a lot painting metal and wooden surfaces. Three huge wooden boxes have been put in length and parallel with the façade. At the outside area, barrels in red and yellow were made into stands that with the mix of the chairs, signs with lights, plants, being under of a huge colourful tent complete the welcome.

Two horses from the carousels welcome you, while they also put limits in the way that the customers move in the restaurant.

Area: 80m² +80m² Out Side Lighting Design: Minas Kosmidis (Architecture in concept) Graphic Design: Minas Kosmidis – Yannis Tokalatsidis Photographer: Studiovd: N.Vavdinoudis - Ch.Dimitriou

Biribildu 在 Basks 的语言里意为……

这是一家快餐店以及与其相关的希腊厨房和相关设计报告。也有存在于可以遥看雅典海滨的 70 平方米的空间内的马戏团和旋转木马的魔法世界的报告。象征马戏团的红黄配色大量出现在金属和木材表面上。三个大木箱的较长方向与外墙方向平行。外部区域内，设计师将红黄相间的桶制成了椅子、有灯标识、植物，共同放置在一个巨大的多色帐篷，整体呈现出温馨友好的氛围。来自旋转木马装置的两匹马欢迎顾客的到来，吸引行人进店一探究竟。

面积：80 平方米＋室外 80 平方米　照明设计：米纳斯·科斯米迪斯（Architecture in concept）　平面设计：米纳斯·科斯米迪斯，扬尼斯·托卡拉特西季斯　摄影：Studiovd 公司 N. 瓦福迪诺迪斯，Ch. 迪米特里

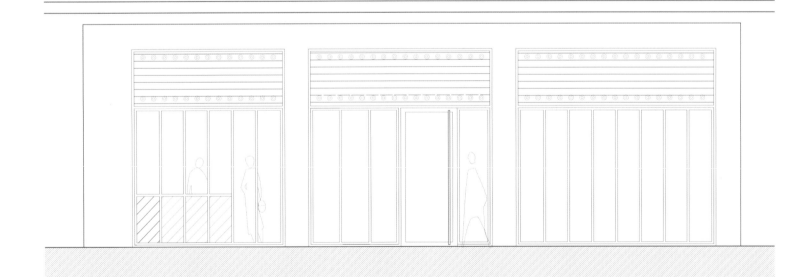

BIASOL: DESIGN STUDIO

HUTCH & CO

Melbourne, Australia

Hutch & Co 餐厅

Biasol: 设计工作室／澳大利亚，墨尔本

Nestled amongst the vines of Victoria's Yarra Valley, 35km northeast of the city is Hutch & Co. From its humble beginnings as the 1800s ironmongery store formally known as Hutchinson's, the site was transformed into a modern restaurant, cafe.

Engaging the full services of the studio, which span interior, building, product and branding design, Hutch & Co was an exercise in composition and contrast between materials, which are wrapped, extruded or intersected to a playful effect. Tasked with delivering a space that considered an architectural extension to the existing building, an outdoor dining area and multi-zoning, the design team decided to strip back and remove all the existing internal walls and linings. With only the exposed brickwork left, they were able to introduce a refined palate of materials and finishes.

This design language extends to the branding, which is bold, recognisable and synonymous with the site's history. Seven meters wide and stretching across the top of the black facade, is the Hutch & Co insignia. Attracting the attention of locals and those passing through the countryside, the frontage echoes the architecture of the local area and was maintained as a design consideration. Set back from the street, this extruding tiled surface area and its built-in bench seating activate the frontage and invite pedestrian interaction.

Throughout the entire design process, the design team were always conscious of the opportunity afforded to them by the Yarra Valley and set out to design a space that continuously captured these views. White tiled walls underpinned by black steel details frame the scenery and lush country terrain, which is referenced internally with greenery from Glass haus. In establishing this level of engagement with the site environs, the design team introduced an element of transparency that meant being able to stand on either side of the building and seeing through to the other side. Opening up the space in this way not only gain natural light and visual access to the breathtaking surrounds, it enabled to achieve a layered effect within the venue, in which the interior and exterior blend to form a single expression.

Completion Date: June 2015 Area: 210m² Photographer: Ari Hatzis Materials: American oak wall panels, American oak timbers, concrete, black steel, white tiles, black tiles

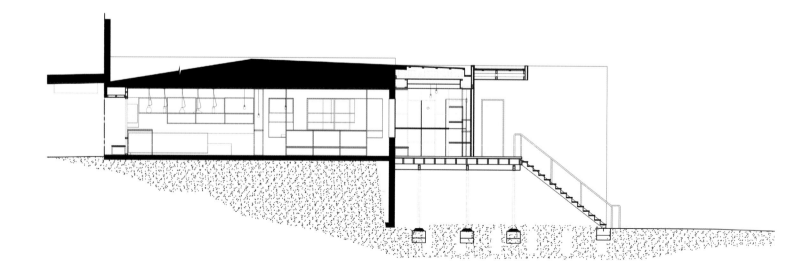

本项目位于城区东北方向35千米的维多利亚的耶那山谷。餐厅创始于19世纪初,最初是一家名为"Hutchinson's"的五金店,如今场地被改造成了现代餐厅和咖啡馆。

项目涉及了设计工作室的全套服务,包括室内设计、建筑设计、产品设计和品牌设计。Hutch & Co餐厅是空间构架与反差的一次尝试。设计师在这些材料上采用或包裹、或挤压、或分割的处理手法,呈现有趣的视觉效果。为了打造一个将建筑延伸至现有建筑的空间设计,设计师安排了一个室外就餐区和多功能分区,将所有原有的内墙和衬里拆除。只剩下裸露的砖块,在此基础上重新加入精致的材料和质感搭配。

这样大胆、辨识度高且与场地历史相符的设计语言延伸到品牌形象设计当中。店铺的标识宽7米,横跨黑色外墙的顶部。吸引本地住户和路人的注意,店面的正面设计呼应当地建筑风格。

整个设计过程中,设计团队一直关注耶那山谷提供的各种可能性,着力打造一个能够持续捕捉这些美景的空间。黑色钢架细节配合白色瓷砖墙壁,将室外的景色和郁郁葱葱的乡村地貌收入眼底。与周边环境建立这种程度的互动的基础上,设计团队还引入了可放置于建筑任意一侧,向另一侧透视的透明元素。以这种方式打开空间格局不仅增加了室内自然光和项目与周围环境的视觉通透度,还实现了街道层面的层次感,室内与室外空间融为一体。

建成时间:2015年6月 面积:210平方米 摄影:阿里·哈齐斯 材料:美国橡木壁板、美国橡木木材、混凝土、黑钢、白色瓷砖、黑色瓷砖

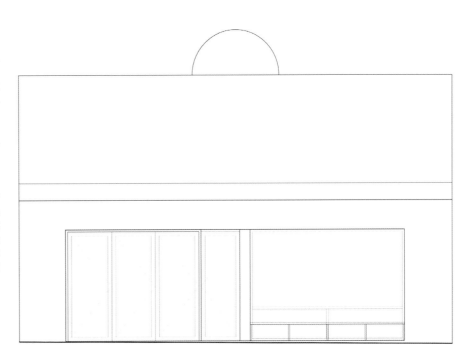

+Mizutaki: an unthickened stew of chicken cooked on the table
注：地鸡火锅是一种未经勾芡处理的在餐桌上现场烹制的炖鸡肉。

TAI_TAI STUDIO

JAPANESE RESTAURANT HAKATA-HANAMIDORI, HARUYOSHI

Fukuoka-city, Japan

博多华味鸟春吉店

tai_tai 工作室 / 日本，福冈

This Japanese 'Mizutak' restaurant is facing the busy roads. This white wall in order to distinguish the store from this road was necessary. The Slits installed in order to light in the store is visible from the outside. The simple white wall that freestanding by being cut from the old building at the slit will be landmark in a complicated place of colour and shape. White wall is painted plasterer. There is a Japanese roof tiles at the top. Sign reminds the SHOJI (paper screens) and paper lanterns. The facade uses collage fragments of old Japanese buildings.

Designer: Wakabayashi Hidekazu   Completion Date: October 2014   Area: 125.4
Photographer: ZIPS create

这间日式"地鸡火锅"餐厅正对繁忙的街道。白色的墙壁将店面与街道区分开来。设计师设置的照明狭缝从店外就看得见。在色彩环境和形状氛围都十分复杂的情况下，简约的白墙与原有建筑的狭缝空间关系具有地标式的设计意义。白墙上粉刷了灰浆。屋顶采用了日式瓦片。标识设计让人联想到纸屏风和纸灯笼。外墙上采用了日本传统建筑元素的拼贴设计。

设计师：荣和若林   建成时间：2014年10月   面积：125.4平方米
摄影：ZIPS create 公司

AKIRA KOYAMA + KEY OPERATION INC.

KOMACHI RESTAURANT

Tokyo, Japan

秋田餐厅

小山光 + KEY OPERATION 设计公司 / 日本，东京

The lot runs along a street lined with disparate buildings. On a street of a commercial district infused with a cluttered ambience, the design team felt that a building could stand out on the street, and possibly bring order to the situation with the construction of a large, unifying façade.

Since exterior stairs have restrictions on their opening ratio, the only option for a single façade that included the staircase was a vertical lattice. They proceeded to select traditional Japanese vertical latticework called tategoshi as an element to render a Japanese visual motif.

A lattice spacing that satisfied the opening requirement for exterior stairs would be too broad for commanding a uniform façade across a large surface. As a solution, the density of the lattice was gradually increased from the front of the staircase across the façade to express unity and variance.

Wood was selected as the material for the lattice to impart warmth, and coated with traditional Japanese red-ochre paint, a natural material used for a long time as a preservative. The Japanese motif was enhanced, and the resulting distinction over the commercial buildings in its surroundings improved the building's presence.

Maximizing floor space on a small lot is contradicted by building codes that inevitably impose setbacks for a building and its staircases. Moreover, many legal restrictions weigh on a building's

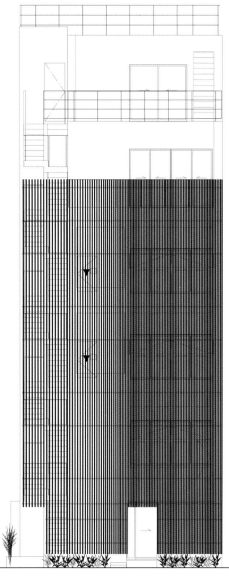

silhouette, such as the installation of balconies considered necessary for securing emergency egress, and result in a disorderly appearance.

However, the messy silhouette of the Komachi Restaurant derived from capturing maximum floor space was painted black to successfully serve like a behind-the-scenes actor to the red-ochre vertical lattice in the starring role.

From inside the restaurant that would mainly be operating from evening-time onward, scenic expectations were poor—possibly a view of the dimly lit tenant building across the street. For such an environment, the design team revisited the building's vertical latticework to consider it as an element that could project characteristic scenes from the interior. Consequently, sections of the vertical lattice were finished as shitomido (traditional Japanese shutters), which could provide scenes like views from a traditional Japanese home with extended eaves. At night, footlights illuminated the shitomido to bring the generated exterior space inside for added spatial depth.

The shutter mechanism allows opening and closing. Since the setback distance can be determined with the shutters fully closed, the volume of the building can be maximized before gaining further volume from the 'eaves' created by opening the shutters. In addition, the shitomido have manual shifters that allow their opening angles to be set freely. Depending on the opening angle of the shutters during the

daytime, the amount of light drawn in can be altered to vary the interior ambience. Any shutter can also be closed for a view through the vertical lattice. By installing shutters to a planar element like a vertical lattice, the façade gains a dimension and can render expressions with light and shadows.

Area: 258.51m² Materials: Wood which was coated traditional Japanese red-ochre paint

项目所在场地位于沿街位置，与街道两旁建筑平行。设计师们认为，在这样一条稍显杂乱的商业区街道上，建筑可以从街道景观中脱颖而出，一个大型的外观统一的建筑甚至可以为街区带来秩序感。

由于外墙楼梯在开放比例上存在限制，只能选择垂直网格形式的有楼梯单侧外墙设计。设计团队进一步选择了名为"馆越"的传统日式垂直网格结构，打造日本风格的视觉主题。

满足外部楼梯对开放度要求的网格间隔如果大面积应用于整体外墙间隔过宽。设计师采取的解决方案是从楼梯正面经过外墙逐渐增加格子的密度，以体现统一和变化的协调统一。

木材为网格增添了温暖的属性，外部涂层为传统日式红赭涂料。它是一种天然材料，长期被用作防腐剂使用。日式风情主题由此得到了强化，相应的商业建筑与周边环境的区别增强了建筑的存在感。

在小面积地块上追求地板空间最大化与建筑规范有所冲突，因而不可避免地需要对建筑本身和楼梯结构加以限制。此外，许多法律限制对建筑的外形有所要求，例如阳台被认为是必要的紧急出口，但会造成建筑外观的不规整。

然而，秋田餐厅由于争取最大占地空间而获得的杂乱的建筑外观被粉刷成了黑色，成功地充当了红赭色垂直网格结构的背景。

餐厅内部主要在傍晚开始营业，因而对周边景色存在的期待不高——夜间能看到的大概只有街道对面灯光微弱的出租楼。针对这样的环境，设计团队再次考察了垂直网格结构，将其作为一个可以将投射室内设计个性特征的元素。对应这一问题，垂直网格中的一些部分采用了日本传统百叶窗的形式，外观看起来就像屋檐加长了的传统日式住宅。夜幕降临时，脚灯照亮百叶窗，将外部空间元素带入室内，增强空间的深度。

百叶窗系统可以打开或关闭。当百叶窗完全关闭时可以得出建筑需要回缩的距离，建筑的体量可以在打开百叶窗形成的"屋檐"之外就得到最大化。另外，这种日式百叶窗配备的手动调节装置可以自由设置百叶窗打开的角度。可以通过控制白天打开百叶窗的角度，控制进入室内的光线，进而调整室内氛围。通过在垂直网格等平面元素上安装百叶窗，外墙的维度得到加强，进行更丰富的光线与阴影表达。

面积：258.51平方米　材料：涂有传统日式红赭涂料的木材

LINEHOUSE

LONE RANGER HOT DOG SHOP

Shanghai, China

独行侠热狗店

LINEHOUSE 建筑设计公司 / 中国，上海

LINEHOUSE was asked to create a new modern fast food identity for Lone Ranger; a hot dog shop, incorporating a wild west theme. Replacing the entire façade of the shop located at the waterfront of the Hangpu River, Shanghai, LINEHOUSE created a playful composition of white timber weatherboards, raw timber, and custom printed tiles featuring an arrowhead motif. The exterior is composed of opening doors and shutter panels, which when opened reveals the continuous tile pattern which wraps the counters and floors. The exterior counters seats 6 with a further 6 seats inside. The 4m high ceiling void is painted bright yellow with a woven rope structure which creates an apparatus for the lights to hang from.

Main Contractor: Jinfang Wang Completion Date: 2014 Area: 19m² Photographer: Benoit Florencon Materials (all locally sourced or custom made): White washed timber weatherboarding rope and black electrical cable, raw timber, custom made white, matt tile with arrowhead motif, powder coated black metal, large Edison Bulbs

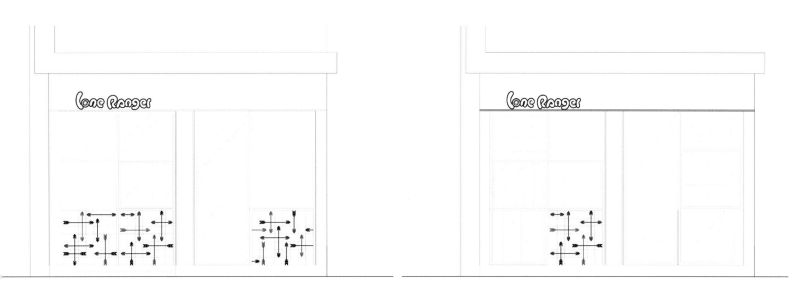

LINEHOUSE 建筑设计公司受邀运用狂野的西部主题为这家名为独行侠的热狗店打造一个全新的现代快餐品牌形象。设计团队将临近黄浦江景的整个店面外墙替换掉，利用白色木质檐板、原木和有箭头图形的白色定制印花瓷砖打造出一个趣意盎然的空间。外墙结构由开放的大门和百叶窗组成，打开时呈现连续的瓷砖图案，将柜台和地板包融在一起。室外座椅可容纳6人，室内可容纳6人。4米高的天花板挑空刷有明亮的黄色涂料，照明装置悬挂在独特的绳索编织设计上。

总承包商：王金芳　建成时间：2014年　面积：19平方米　摄影：伯努瓦·弗洛伦康　材料：（所有材料均为本地生产或定制）白色过水木质檐板、绳子和黑色电线、原木、有箭头图形的白色定制亚光瓷砖、有粉末涂层的黑色金属、大号螺口灯泡

MEI ARCHITECTS AND PLANNERS

SHINING MCDONALD'S IN HEART OF ROTTERDAM

Rotterdam, the Netherlands

鹿特丹市中心的闪亮麦当劳

Mei 建筑规划公司 / 荷兰、鹿特丹

The new building volume has been carefully detailed and articulated by Mei to open up views of the monumental post office behind it. As a result, the pavilion has the most compact possible core, with glazed facades all around. A fully transparent lobby, with entrances on three sides, makes it seem as though the public space flows through the building.

To strengthen the connection between the pavilion and its surroundings, the outdoor terrace will feature the same furniture found in other public spaces in the city of Rotterdam.

Like the historical buildings in the area, the pavilion features a richly articulated facade, carried out in one single material: gold anodized aluminum. This warm and elegant-looking material is vandal-proof and enduring at the same time.

As McDonald's is open day and night (24/7), its appearance after dark is important. By day the building is inviting to shoppers, while in the evening it glows to attract the nightlife crowd. The skin is perforated with heart-shaped openings to form a 'veil' around the glazed building through which illumination shines. This skin is continued in the interior walls and ceilings. With the application of various degrees of perforation, the facade depicts a crowd of people on Coolsingel. After all, the new McDonald's is built for the people of Rotterdam, who now have yet another reason to be proud of their city.

The various power boxes for utilities and traffic regulation systems, always an eyesore on the street, are integrated into the building. Moreover, the incorporation of illumination into the façade will enhance safety on the street. Technical installations are integrated within the roof, which is designed as a fifth facade.

Structural Engineer: Adviesbureau Roelen, Eersel Façade Designer: Glasimpex, Schiedam; VPT Versteeg , Heusden Photographer: Jeroen Musch, Rotterdam; Ossip van Duivenbode,Rotterdam; Frank Hanswijk, Rotterdam

新建筑体量的细节精致丰富，通过 Mei 建筑规划公司的方案打开建筑后方邮局的视野。餐厅的核心部分因此相当紧凑，四个方向都使用了玻璃墙面。入口分布在全透明大厅的三个方向，使得公共空间似乎在建筑内自由地流动起来。为了强化小亭和周围环境之间的联系，室外露台上将会使用鹿特丹城区的其他公共空间内使用的相同家具。与项目所在区域内的历史建筑类似，餐厅的外墙通过单一的材料——阳极镀金铝——表现出丰富的设计思考。这种温馨而优雅的材料不仅能够防破坏，也具备耐磨的特点。

鉴于麦当劳昼夜开放(24/7)，店面在天黑以后的外观就显得十分重要。白天，店面吸引的主要是购物人群，夜间的主要受众则是夜生活活跃的人们。外墙上的心形穿孔在玻璃墙壁周围形成一层"面纱"，灯光透过它散发光芒。这一元素一直延续到室内墙壁和天花板上。设计师通过外墙上不同大小的孔隙描绘了库尔辛格大街上的人群。毕竟这家新麦当劳是为鹿特丹人而建的，而鹿特丹人又有了另一个为城市骄傲的理由。公共事业和交通系统的各种电源盒通常是街道上一处碍眼的事物，本案中被结合到了建筑中。融合在外墙设计中的照明技术提升了街道的安全指数。技术设施安装在了屋顶，作为建筑的第五、第六层。

结构工程：艾德维斯比罗·勒伦设计公司，埃尔瑟尔　外墙设计：Glasimpex 设计公司，斯希丹，VPT Versteeg 设计公司，赫斯登　摄影：杰隆·姆驰（鹿特丹），奥西普·对文博德（鹿特丹）

HISANORI BAN KAZUMOTO TERASHIMA

MIRRORS

Gifu City, Japan

镜子餐厅

Hisanori Ban Kazumoto Terashima 设计公司 / 日本，岐阜市

A row of cherry trees is planted at an embankment at its basin, and many people visit this location during the cherry-blossom viewing season. There is Mirrors along the avenue. Taking advantage of this location, the designers intentionally made repeated refractions of the tree. In order to amplify the cherry two mirror walls set up angle position and making a cherry forest on a corner of the town. Overlapping the cherries and reflecting with a warp, the people are invited by it to the forest. In addition, in the café the people could see the cherries and the reflecting cherries at the same time and feel season changing closely.

Especially vertical roof struts of motif of the tree blanch made them feeling of a rest under the tree. Between two buildings, white gravel makes more reflecting brightly and symbolically expressing the center of a tree. There are three trees in this site, and the trees are camellia. Camellia makes flowers before cherry blossom. First, camellia has red flowers, second cherry has pink flowers. The team designed changing season of winter to spring by the colours. Exterior wall is white steel that is galbanum plating. Entrance door is red. Interior wall is red and green. Every design and colour implies being in the forest intentionally.

Lighting Design: Atsuko Fujita / Koizumi Lighting technology Area: 99.07m² Photographer: Shigetomo Mizuno Materials: exterior / roof and wall - stainless steel, galbanum steel

这块盆地的路堤边种了一排樱桃树,许多人在樱花开放的观赏季节来到这里。本案就坐落在这样一条大道上。

设计团队充分利用地理优势,有意地对树木进行多次折射。为了扩大樱桃树的视觉效果,竖立起两面镜面墙,墙壁互成角度,在城市一角打造出樱桃树的森林。重叠的樱桃树影像加上变形效果吸引人们探索这片森林。另外,人们在咖啡厅里也可以看到樱桃树和樱桃树的影像,密切感受季节的变迁。

两栋建筑之间的白色砾石使得反射效果更为明亮,也象征性地表现出树木的核心所在。场地中有三棵树,是山茶花。山茶花的花期比樱花早,而且山茶花的花为红色,樱花是粉色。冬季到春季的设计也是按色彩进行的。

外墙采用的是白色阿魏脂镀层钢材。入口大门是红色的,室内墙壁则是红绿相间。每个设计和配色都显示出森林的寓意。

照明设计:藤田敦子 / 小泉照明科技公司　面积:99.07 平方米
摄影:水野重友　材料:外部 / 屋顶和墙壁——不锈钢、阿魏脂钢

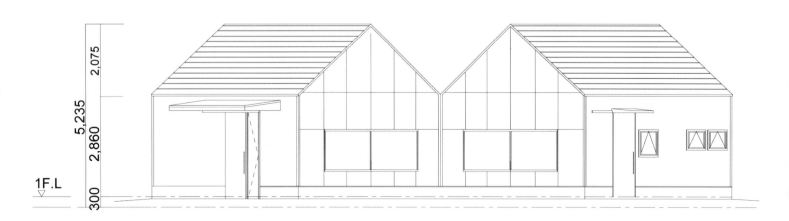

YOSHIHIRO KATO ATELIER CO., LTD.

MOTSUNABE SHINRAKAN

Nogoya-shi, Aichi, Japan

新罗下水火锅餐厅

加藤义弘工作室有限公司 / 日本，爱知县，名古屋

'Shinrakan' that was opened in an area crowded with offices and condominiums has a rectangular shape, five meters wide and twenty-five meters long. The shopfront sign is used urethane baking coated aluminium plate t2.0 and LED indirect illumination, and also the letter signboard is used acrylic box type letter sign. External wall is siding board t12 covered with urethane baking coated aluminium plate t2.0, and the window is used window: urethane baking coated steel frame grass t6. The shopfront of this restaurant creates an atmosphere of silence, the sign and exterior walls are all painted in black, especially at night if you do not pay a special attention to this restaurant, people probably will miss this fabulous restaurant. Though black as the main color of this restaurant, various materials and layouts make black into several shadows. It can give customers a fresh impression with 'forms' rather than depending on special techniques or too much decoration. The lightings were also made originally for the restaurant with various illuminations, and it expresses some movements and flows, just like its interior design in the space.

Designer: Yoshihiro Kato   Completion Date: October 2014   Area: 119.90m²   Photographer: Nacása & Partners Inc.

这间餐厅开设在办公和公寓密集区域。餐厅呈长方形，宽 5 米，长 25 米。店面标识采用 2.0 聚氨酯烘烤涂铝板和 LED 间接照明装置，字母布告板使用丙烯酸箱式字母标识。外墙是厚度为 12 毫米的聚氨酯涂层铝质墙板，窗户使用厚度为 6 毫米的聚氨酯涂层钢框架。尽管黑色是餐厅店面的主色调，不同材料和布局的应用使得黑色呈现不同深度。店面营造出静谧的气氛，标识和外墙也都粉刷成了黑色，这样的形式并不依靠过多的装饰，也会在夜间给顾客留下新奇的印象。各式各样的照明装置也全部是为餐厅特别设计的，表现流畅的动感，与其他室内设计元素相辅相成。

设计师：加藤义弘　建成时间：2014 年 10 月　面积：119.90 平方米
摄影：Nacása 合作公司

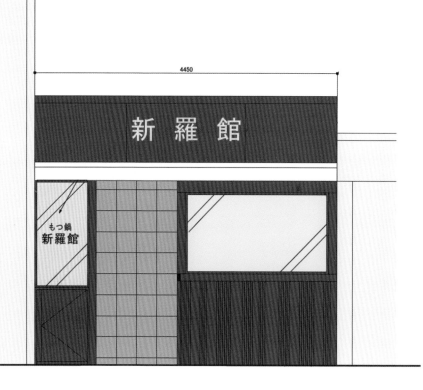

CUT ARCHITECTURES

PNY OBERKAMPF

Paris, France

PNY Oberkampf 餐厅

CUT 建筑事务所 / 法国，巴黎

Following the success of the first Paris New-York restaurant opened in December 2013 Rue du Faubourg St Denis in the 10th district and also designed by CUT architectures, PNY Oberkampf claims it difference while taking on its heritage.

Facing the renown Nouveau Casino club, the second venue offers a more urban and alternative image than the first one. The facade is kept untouched though it's black spray-painted all over while the sign is put up in the middle of it: P,N,Y, three thin neon tube letters set in raw aluminum frames.

The floor is in cement tiles suggesting the continuity of the pavement within the restaurant space.

Area: 80m² Completion Date: April 2014 Photos: ©David Foessel

2013 年 12 月开始营业的巴黎－纽约首家餐厅获得成功之后，PNY Oberkampf 餐饮公司继续邀请 CUT architectures 设计公司在第二家店面中展现不同的品牌面貌，同时体现了品牌精髓。

新店面正对着著名的巴黎新赌城俱乐部，与第一家店面相比，这里提供更为城市化的环境。店面外墙保持不变，整体喷成黑色，店面标识悬挂在中央。P，N，Y——三个狭窄的霓虹灯管分布在铝框架中。

地面铺设的是水泥瓷砖，将人行道的经典元素延续到餐厅空间里。

面积：80 平方米　建成时间：2014 年 4 月　摄影：大卫·福塞尔

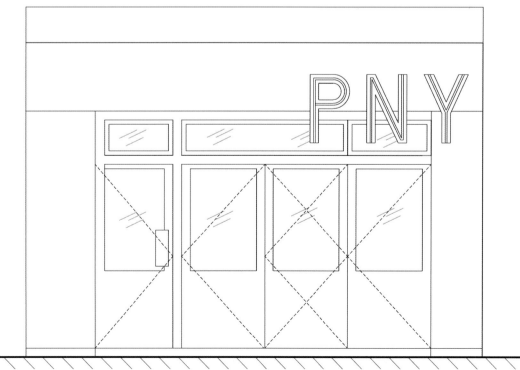

BEDMAR & SHI

RESTAURANT THE PAPAGAYO

Cordoba, Argentina

帕帕加约餐厅

Bedmar & Shi 设计公司 / 阿根廷,科尔多瓦

'El Hueco' (The Hole), as the designer uses to call it, is just 2.40 meters wide and 32 meters long, with a height of almost 7 meters. Originally it had a reinforced concrete slab, very few natural light and two brick walls from 1870. The idea was to create a place full of natural light so Bedmar & Shi replaced the slab with a glass emphasizing the interior height. The facade was designed in a very simple way, also emphasizing the height of the building. The required spaces were toilets, the kitchen, the dining room, a private room and an office upstairs. All these required the installation of some technical elements such as cables, pipes and ventilation. So it is decided to use one of the walls, which is now finished in exposed concrete. The two walls now represent the contrast between the contemporary and the antique. This determined the location of the cellar and toilets, which have timber cladding, separated from the slab, creating a small building inside another.

Designer: Ernesto Bedmar Completion Date: 2015 Area: 120m²
Photographer: Gonzalo Viramonte

本案中，设计师将原有建筑称为"洞"，因为它仅有 2.4 米宽，32 米长，高度接近 7 米。原有建筑结构包含一块钢筋混凝土板，两面 1870 年建造至今的砖墙，和很少的自然光照。设计团队希望打造一个自然光照充足的空间，所以用玻璃替换了混凝土板，突出楼体的高度。外墙的设计十分简单，对楼体的高度做进一步强调。项目空间包括卫生间、厨房、餐厅、私人房间和楼上的办公室。这就要求安装一些技术元素，例如电缆、管道和通风设施。于是，设计师决定利用其中一面原有墙壁，结合裸露混凝土表面。与另一面墙壁形成传统与现代的对比。这也决定了地窖和卫生间的位置，同时使用木材覆面，形成建筑中的小建筑。

设计师：埃内斯托·贝德玛尔　建成时间：2015 年　面积：120 平方米　摄影：冈萨洛·维拉莫特

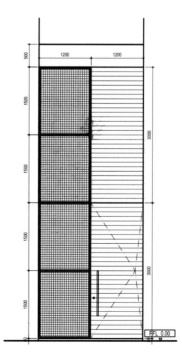

VINCENT LOEFF, BURO BLASÉ LOEFF

ROOST

Amsterdam, the Netherlands

Roost 咖啡店

文森特·勒夫，布罗·布拉泽·勒夫 / 荷兰，阿姆斯特丹

This former bookshop has been transformed into a coffee take-away offering breakfast, lunch and cakes. Situated in the eastern part of Amsterdam opposite of the 'Onze Lieve Vrouwen Gasthuis' hospital, Roost plays a central role in catering to the neighbouring residents, nurses and passers-by. All of the products sold in Roost are homemade by the owner, from soup, salads, croissants and sourdough buns to all the cakes. Specialty coffees are prepared with either a state-of-art espresso machine, or according to various 'slow-coffee' methods according to the customer's order.

The shop was completely stripped to its bare essentials revealing the original concrete structure and then built from the ground up to create a welcoming cozy venue. Central in the lay-out is the counter and open kitchen where customers can see first-hand the preparation of all the products sold in Roost. The window seats make for intimate spots and also function as extra storage and product display.

Area: 100m² Photographer: Jos Kraaijeveld

项目所在地曾经是一家书店，如今经过改造变成了一间咖啡外卖店，供应早餐、午餐和蛋糕。

店面位于阿姆斯特丹东部，对面是"Onze Lieve Vrouwen Gasthuis"医院。咖啡店在为社区居民、护士和路人提供饮食方面发挥了重要作用。店内出售的所有商品，从汤、沙拉、羊角面包和酵母面包再到各种蛋糕均为手工制作。招牌咖啡使用先进的意式咖啡机制作，或者根据顾客要求采用"漫咖啡"手法制成。

设计师完全摒弃了店面原有的外墙，露出裸混凝土结构，从头开始打造一个温馨舒适的场所。顾客们可以在位于布局中央位置的柜台和开放式厨房亲眼看到所有食材的处理烹饪过程。靠窗的位置适合亲密的约会，也可以作为额外的储存空间和产品展示功能使用。

面积：100平方米　摄影：乔斯·克莱杰维德

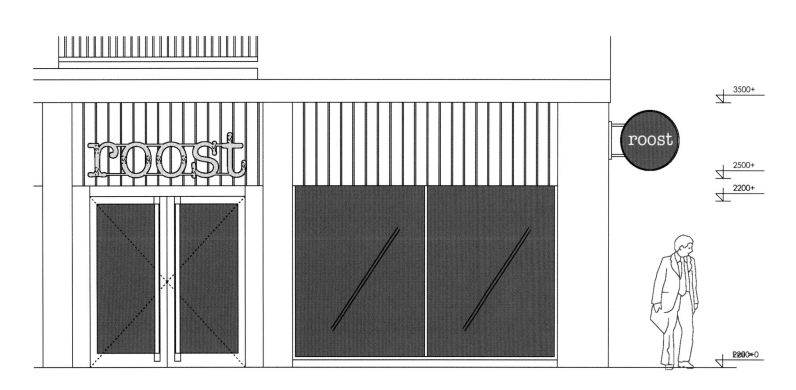

TALLER KEN

SAÚL BISTRO, PRADERA CONCEPCION

Guatemala City, Guatemala

"索尔酒馆,拉德拉·康塞普西翁"

Taller KEN 设计公司 / 危地马拉,危地马拉城

'Saúl Bistro Pradera Concepcion' received an American Institute of Architects 2015 restaurant design award. For 'Saúl Bistro Pradera Concepcion' the designer team wanted to challenge this typology of the generic suburban shopping center. Because the project is located at a high elevation, on the road to El Salvador, the client wanted the cozy feeling of a mountain house. To achieve this, the project needed to be separated from the commercial center to no appear like a theme-restaurant.

As a result, red corrugated house-shaped containers cantilever out past the usual infill constraints of the shopping mall, giving the project the eye-catching feeling of an art installation or something that is extra-ordinary from the context. This subversive approach to the site gives the clients a stronger commercial presence and questions the conventional building type. This attitude largely contrasts attitudes for the region regarding retail and commercial projects. In order to create intimate spaces each container is placed within the existing concrete structural grid of the mall.

Completion Date: 2014 Area: 450m² Photographer: Andres Asturias

这个名为"索尔酒馆,拉德拉·康塞普西翁"的项目获得了美国建筑师学会2015年度餐厅设计大奖。设计团队希望在项目中挑战郊区购物中心的类型设计项目。由于项目所在位置较高,且位于去往萨尔瓦多的路上,委托方希望打造出如山间小屋般舒适的氛围。为了实现这个目标,项目需要与购物中心分离开,避免给人造成主题餐厅的印象。

因此,房屋形状的红色波纹容器从购物中心结构上支出,为项目赋予了引人注目的艺术感和突出感。这种颠覆性的场地处理方法为委托方争取了更强的商业存在感,同时也对传统的建筑类型提出了质疑和挑战。这种态度与零售和商业项目的处理态度形成了鲜明的对比。每个结构单元都放置在商场的混凝土结构内部,以便打造出更为私密的空间。

建成时间: 2014年　面积: 450平方米　摄影: 安德烈斯·阿斯图里亚斯

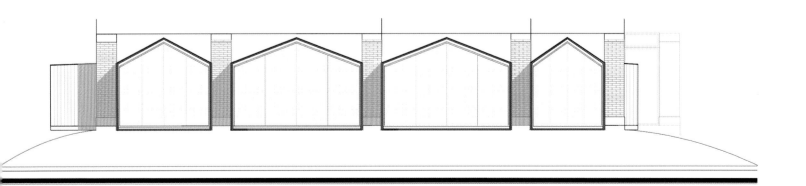

TALLER KEN

SAUL E. MENDEZ, ZONA 14

Guatemala City, Guatemala

14 区 Saul E. Mendez 餐厅

Taller KEN 设计公司 / 危地马拉，危地马拉城

'Saul Zona 14' is a house with many charms. Under one roof, the designers combine fashion, design, art, objects and food. Here, a union of culture and commerce full of surprises and small treasures entices clients and visitors to take the time to enjoy diverse experiences. It is a place to meet, to stop for a drink at the café, to shop at the store or to visit the terrace to have a long meal or just sit to read. The design's playful qualities draw from cultural references. The buildings the façade draws inspiration from surrounding architectural heritage of Guatemala City- namely the deep ornamental window niches of the preserved Spanish colonial. The buildings exterior surface is covered in molded fiberglass panels with a plaster stucco finish to achieve the smooth and stretched quality. The combined effects give the sensation the window is 'actively' pushing it out from the wall. This has the pragmatic benefit of freeing valuable interior space and creating window-box vitrines for display. The exterior terrace features a colourful canopy, made of 1,000 pounds of thread hanging from a steel structure, has tactile qualities of softness. It has a visually vibrant palette of green and yellow blends enhancing the surrounding vegetation. The source of inspiration was the natural production techniques still used by indigenous people in Guatemala.

The drying process of hanging treads is part of highly traditional elements of craft culture. Ultimately this seemingly decorative element is highly pragmatic; acting as both a solar shade and sound absorbing surface which make the space cool and intimate.

Area: 450m² Photographer: Andres Asturias

"14区 Saul E. Mendez 餐厅"是一个充满魅力的地方。设计师在它的屋檐下融合了时尚、设计、艺术、物件与美食。这个充满惊喜和小玩意的文化和商业的结合体吸引顾客和游人花时间享受五彩纷呈的体验。人们可以在这里聚会，驻足享受一杯美味的饮品，或者尽情购物，悠闲地用餐，或者只是静静地读书。这里趣味无限的设计在当地文化中吸取灵感。

店面外墙的设计灵感来自周围的危地马拉城历史建筑——深陷的装饰窗户保留了西班牙殖民时期的特点。建筑外部表面覆盖模塑玻璃纤维板，抹灰表面展现光滑、伸展的特点。二者结合呈现出的效果赋予了窗户"主动"从墙面上进出的动感。这样的设计有一个实在的好处，即增大了宝贵的室内空间，也打造出适合展示商品的橱窗。

室外平台上有一面色彩斑斓的帆布篷，它由1000磅的线悬挂在钢结构上，具有柔软的质感。

面积：450 平方米 摄影：安德烈斯·阿斯图里亚斯

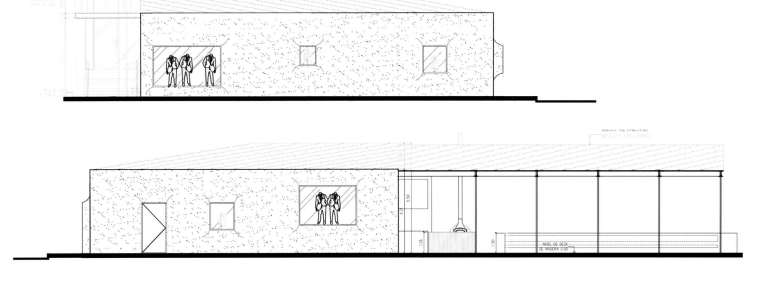

YOUSEF MADANAT ARCHITECTURAL STUDIO

SEKRAB RESTAURANT AND BAR

Amman, Jordan

Sekrab 餐厅与酒吧

优素福·马达纳特建筑设计工作室 / 约旦，安曼

Sekrab is the Arabic slang for Scrap Yard, and as the name implies, the entire façade was made out of discarded metal sheets, scrap yard elements and totaled vehicles. The main idea behind this design was to do something unique that was never attempted in the region before, and to shed light on the importance of up-cycling to the environment, and that there actually is beauty in trash. The interiors follow the same concept as all furniture used are configurations of discarded materials.

Designer: Yousef Madanat Completion Date: 2014 Photographer: Aladdin Qattouri

"Sekrab"一词在阿拉伯语中是废料场的意思。正如店名所述，整个外墙使用废弃金属板、废料场元素和废弃车辆制成。打造一个独特的、在该地区前所未见的项目，宣扬环境升级改造的重要性以及发现垃圾中的美是设计背后的主要思想。室内设计遵循相同理念，项目中使用的所有家具都采用废弃材料制成。

设计师：优素福·马达纳特　建成时间：2014年　摄影：阿拉丁·加特托里

DENYS & VON AREND

SENSACIONS CATERING

Barcelona, Spain

感官餐厅

Denys & von Arend 设计公司 / 西班牙，巴塞罗那

Whenever you try to create a brand from scratch, to corporatize, beautify or modernize a concept traditionally associated to low costs and immediacy, as it is the case with take away food retail, there's always the latent risk of snatching at the same time the personal aura, the crafty feeling and the accesible appeal this type of commercial establishments usually have. Therefore, the operation of transforming a former basement garage, in downtown Sabadell, into a catwalk for 'Haute Cuisine Pret-a-Porter' was evidently raised from the very beginning as a major challenge that needed to harmonize all contradictions that such merger of concepts arose, while facade and overall branding needed to fullfill the needs of two quite distinct activities, address two different audiences, and promote, simultaneously, two ways of consuming high-class catering and low cost gourmet dinners. The black walls seem to want to go unnoticed. Colour and graphics in the shop resemble the corporate image system designed for the brand, and this in turn is closely related to the aesthetics of the place, since they were born in parallel. This symbiosis becomes evident in the facade, which combines the graphic pattern of the tablecloth, the logotype, the coporate 'values' (instead of colours), the chipboard chip, the transparency of glass and specific lighting as a prelude to an experience which aims to stimulate all the senses.

Designer: Dani Pérez y Felipe Araujo   Completion Date: 2014
Area: 150m²   Photographer: Victor Hugo Anton

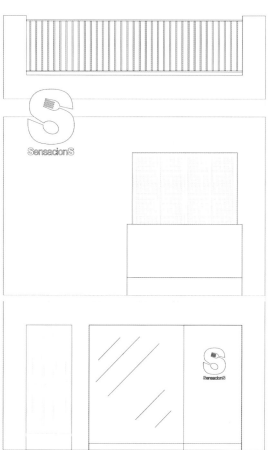

无论何时，想从头打造一个品牌，在外卖食品零售业中将一个传统意义上低成本、易操作的概念企业化、美化或现代化。这类商业项目常常存在抢夺商机的潜在风险，也受个人气场、产品工艺和亲和力等因素影响。因此，将位于萨瓦德尔市中心的前地下车库改造成"高级料理成衣"的秀场这一想法从一开始就是一个巨大的挑战，需要协调所有存在的矛盾方面，同时外墙设计和整体品牌形象必须满足两个截然不同的活动需要，面向两种不同的受众，推广高端餐饮和低价美食两条路线。黑色墙壁非常沉稳低调。店内的色彩和图案设计与企业的品牌形象系统十分相近，进而使得企业品牌形象得到强化。这种共生关系在外墙设计上尤为明显，结合了桌布图案、品牌标识、企业"价值"（而非配色）、纸板碎片、透明玻璃和特制照明装置，激发所有的感官。

设计师：达尼·佩雷斯，费利佩·阿劳霍　建成时间：2014 年　面积：150 平方米　摄影：维克多·雨果·安东

ATELIER ALTER

YINGLIANG STONE RESTAURANT

Beijing, China

英良石餐厅

时境建筑设计事务所 / 中国，北京

This project transforms a deserted factory plant into the supporting facility of an exhibition building of small stone materials. The owner wishes to create a place for designers to exchange ideas.

Stone is one of the earliest construction materials exploited by human. It traditionally represents authenticity, solemnity and craftsmanship while with the help of modern processing methods, stone materials become lighter and thinner, at the same time much easier to install, losing its original strength and materiality.

In this restaurant project, designers further explore the nature of this material and challenge the weight and solidness of stone to present a brand new experience of this material and the space as a whole.

Completion Date: 2016 Leading Designer: Bu Xiaojun, Zhang Jiyuan Design Team: Qin Kai, Li Zhenwei, Du Dehu, Liu Tongwei Area: 110.65m² Image Copyright: Atelier Alter

127 • 餐厅

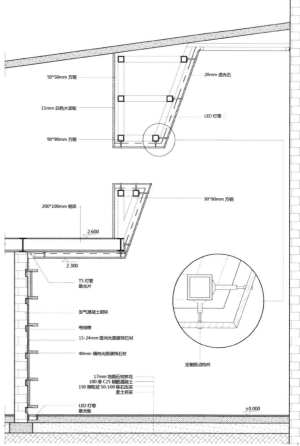

这是一个废旧厂房改建项目，作为一个小型的石材展示馆的配套设施，石材商业主想通过这个改造项目提供一个与设计师交流的场所。

石材是人类最早最广泛使用的建筑材料之一，人类使用石材的历史和开采、加工工艺既古老又崭新。石材作为建筑材料本身往往代表着真实、庄重、工匠精神，而在现代工业新的加工方法中，石材变得愈加轻薄、光洁以迎合轻便、整洁的安装需求，石材在这个逐渐二维化的转变中慢慢失去了其本身的力量感和精神性，从而失去其原本的物质性。

在这个餐厅的设计里，我们试图深入到材料的性质本身，挑战石材的重量感和坚实感，制造新的人们对物体本身的感受，通过这种不寻常影响人们对空间的体验。

时间：2016年　主要设计师：卜骁骏　张继元　设计团队：覃凯、李振伟、杜德虎、刘同伟　面积：110.65平方米　照片版权：时境建筑设计事务所

BRANDON BRANDING AGENCY

SIMPLE

Kiev, Ukraine

简单餐厅

布兰登品牌设计公司 / 乌克兰，基辅

The idea of the restaurant is 'be simple, eat simple', it implies cooking from local, fresh, not preserved products, but in unusual combinations. For this reason, the designers used natural colours and simple materials like wood, plywood, craft paper, etc. without complicated refinement. The creative team also decided to look at usual details from a different point of view. That's how they got a shovel as a door-handle, rakes as coat hooks, rolling pins as a menu for drinks, concrete lamps made of recycled plastic bottles and so on. They honestly bought all that at the market and adjusted in the interior. Posters are the food for guest's eyes - the subject of study and interaction. The designers decided to tell a great story that goes as a cross-cutting theme through all posters. It's a story about how something simple turns complex, it focuses on the main idea of the restaurant: making interesting and tasty dishes out of basic/local products. Each poster is supported by a detailed commentary that can tell you, for example, about languorous cow Agnessa, or how to make dough that will speak.

Designer, illustrator: Olga Novikova Designer: Elena Parhisenko, Anton Storozhev Visualization: Nikolay Mihov Completion Date: October 2014 Photographer: Andrey Shalimov, Yana Korobenko

这家餐厅的设计理念是"简单做人,简单进食",以不同寻常的方式传递用地产的新鲜食材烹饪的美食理念。为了更好地表达这一理念,设计师使用天然配色和木材、胶合板、牛皮纸等简单材料,以简约得体的形式打造用餐空间。设计师还用不同的视角处理寻常的细节,所以我们见到了用作门把手的铲子,用作衣钩的耙子,用作饮品单的擀面杖,以及回收塑料瓶制成的混凝土灯等。设计师在市场购入所有材料,根据室内需求做出调整。对顾客来说,欣赏店内海报设计也是一种视觉享受——它们是学习和互动的主题。设计师决定通过所有海报讲述一个精彩的故事。讲述简单的事物是如何复杂化的,关注餐厅的主题:用基本/当地食材制作出有趣且美味的菜肴。每张海报配有详细的说明,向观众讲述怠惰的奶牛艾格尼萨的故事,或者如何制作一个会说话的面团。

设计师/插画师:奥尔加·诺维科娃 设计师:埃琳娜·帕里先科,安东·斯托罗哲夫 视觉设计:尼古拉·米霍夫 建成时间:2014年10月 摄影:安德烈·沙利莫夫,雅娜·克罗本科

TALLER DAVID DANA ARQUITECTURA
MERCADO MORERA

Mexico City, Mexico

梅尔卡多·莫累拉餐厅

塔莱·大卫·达纳建筑师事务所 / 墨西哥，墨西哥城

Mexico D.F. is developing new and extraordinary urban interventions with the aim of utilizing the existing cavities with potential. In this case the 'Mercado Morera' located below one of the most congested freeways becomes one of the great examples of this ongoing scene. Yumi-Yumi is a brand new concept for Japanese cuisine; the client asked to design an integral project including the brand identity. The design proposal had the objective of being easily replicated in multiple locations. TDDA partnered with Diego Leyva, Creative Director of Nhomada to develop the graphic design, they worked through a multidisciplinary process to engage the industrial design, typographies and architecture. The result transforms the typical sushi bar and integrates a whole concept for the traditional rice bowl.

Area: 12m² Designer: David Dana Cohen, Juan Castañeda Photographer: Alessandro Bo

墨西哥城正在开发全新的城市干预项目，目的是利用具有潜力的现有空间。位于最拥堵的高速公路下方的"梅尔卡多·莫累拉餐厅"成了这一项目中最典型的一例。委托方诚挚邀请塔莱·大卫·达纳建筑师事务所（下称TDDA建筑事务所）为店面进行品牌形象设计。

设计方案需要易于在多个地点进行复制。TDDA建筑事务所与Nhomada设计公司的创意总监迭戈·莱瓦合作完成了项目的平面设计，多学科的工作过程涉及工业设计、排版和建筑设计。

面积：12平方米 设计师：大卫·达纳·科恩，胡安·卡斯塔涅达 摄影：亚历桑德罗·波

CHIOCO DESIGN

TORCHY'S TACOS – SOUTH CONGRESS

Austin, USA

南国会大道 Torchy's Tacos 餐厅

Chioco 设计工作室 / 美国，奥斯汀

Consciously referencing the archetypal 50's-era drive-in diners, Chioco Design developed an iconographic structural 'X' & 'Y' form traversing the length of the block facing South Congress Ave. These bright red columns support a crenellated roof profile extruded through the entire building, seamlessly connecting inside to out, and allowing for the addition of numerous north-facing skylights which provide an abundance of consistent natural light. Steel planters and tree wells extend out to the sidewalk and street to act as an inviting and accessible area adjacent to the vibrant energy of South Congress Ave. Encaustic cement tiles are thoughtfully used around the interior and exterior. Brightly coloured custom fabricated light fixtures hover above both bar areas serving as a significant contrast to the white ceilings during the day and brilliantly illuminating the space at night.

Completion Date: January 2016 Area: 268.21m² Photographer: Patrick Y Wong/ Atelier Wong

以 20 世纪 50 年代汽车餐厅为灵感，Chioco 设计工作室打造了图形化的"X"和"Y"形设计，横贯面向国会大道的整个方向。这些亮红色的支柱支撑起屋顶结构，联接了室内外空间。此外，多个北向天窗为室内提供充足的自然光。钢质花盆和树井向人行道和街道延伸，店面内外巧妙地使用了釉面水泥瓷砖。色彩鲜艳的定制照明装置悬挂在吧台上方，白天与白色天花板形成反差，夜晚照亮整个店内空间。

建成时间：2016 年 1 月　面积：268.21 平方米　摄影：帕特里克·Y·黄 / Atelier Wong 工作室

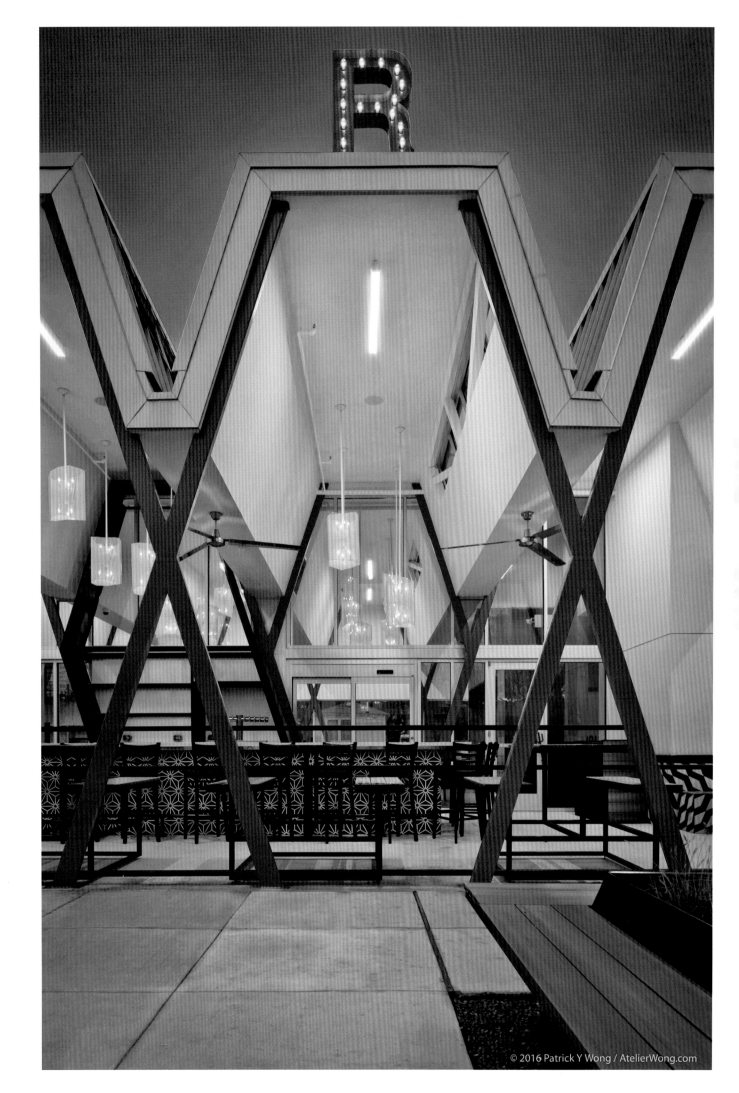

© 2016 Patrick Y Wong / AtelierWong.com

SUPERCAKE SRL

MANTRA RAW VEGAN. MARKET, RESTAURANT AND STATE OF MIND

Milan, Italy

曼特罗生食素食餐厅

Supercake Srl 公司 / 意大利，米兰

The architectural, graphic design and communication standards for Mantra came out of the idea of a seed and of essence. Simplicity, then, became the guiding principle for the whole project. It was pursued in the architectural and graphic design, their form and materials. The aim was to deprive everything of unnecessary frills and encourage rigour, harmony and functionality.

Market, restaurant and state of mind.

'Anyone who enters is welcome, regardless from its usual diet.

Who comes out takes away a smile, an experience of purification, a seed of rebirth.'

Construction Design Team: Mediterranea Costruzioni Srl
Completion Date: 2014 Area: 130m² Photographer: Valerio Gavana

曼特罗餐厅的建筑设计、平面设计和通信标准来自种子的概念和事物的本质。简洁是整个项目的指导原则。建筑设计和平面设计，无论材料还是形式，追求的也是这样的理念。其目的是去除所有不必要的装饰，严谨、和谐和功能。

市场、餐厅和心境
"这里的大门向所有人敞开，无论其日常饮食习惯如何。
来到这里的人离开时会带走笑容，净化的体验和重生的种子。"

建筑设计团队：Mediterranea Costruzioni Srl公司　建成时间：2014年　面积：130平方米　摄影：瓦莱里奥·加瓦纳

CUT ARCHITECTURES

CAFE COUTUME AOYAMA

Tokyo, Japan

青山 CAFE COUTUME 咖啡店

CUT 建筑师事务所 / 日本，东京

The aim was to keep the strong identity designers created for the Coutume brand while adapting it to the Tokyo location. At the crossing between a Parisian coffee shop and a laboratory, Café Coutume Aoyama offers a two sided space: On the entrance side the laboratory is set-up under a white hygienic grid ceiling with integrated LED panels lighting up the bar. The wall base of the interior walls as well as on the exterior facade is clad with white tiles up to 1m creating a continuous line surrounding the entire space both inside and outside. On the street side another block made of tiles and glass is hosting a roasting sampler, high stools and small Japanese plants. Behind the bar, a translucent glazed wall hiding the kitchen integrates glass shelves for display.

Completion Date: April 2014 Area: 85m² Photographer: ©David Foessel

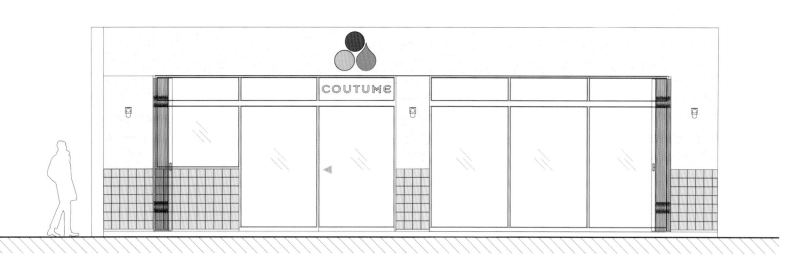

本案的设计目标是在保留强烈的品牌形象的前提下,为东京分店打造适合当地环境的店面。

青山 CAFE COUTUME 咖啡店是巴黎式咖啡店和实验室的综合体,向顾客呈现一个两面空间:入口一侧的实验室白色格栅天花板下设置实验室风格空间,集成 LED 面板照亮了酒吧区。

室内墙壁的墙基以及室外墙壁铺设白色瓷砖到 1 米高度,在整个室内外空间创造出连续的统一线条。

街道一侧,瓷砖和玻璃组成的另一处空间内可以见到食品加工台、高脚凳和小型日本植物。吧台后方是半透明的玻璃墙,将厨房巧妙地隐藏起来,同时构成了用于展示的玻璃货架。

建成时间:2014 年 4 月　面积:85 平方米　摄影:大卫·福索尔

MAURICE MENTJENS

COFFEESHOPS SKUNK AND RELAX, SITTARD

The Netherlands

锡塔德咖啡店

莫里斯·蒙切斯设计公司 / 荷兰

To comply with the stringent rules and regulations, Hendriks wanted two entirely separate coffeeshops in one building. Each had to have its own entrance and exit, although an internal connection was allowed. Combined with a rather inconvenient triangular site, these requirements produced an unusual floor plan. 'Takeaway' outlet Skunk dominates the view from the street, with passageways on either side leading to Relax, which is tucked away at the rear of the building. Those two access routes take up a substantial amount of space, cutting the overall retail floor area despite the two outlets sharing facilities like toilets, a service area and a security booth. 'We have turned that drawback into an advantage by transforming the passageways into spectacular experiential spaces,' says Mentjens. The secret lies in the panels lining them, which are mirrored along the outside wall and transparent on the inward-facing side. The resulting effect makes walk-in coffeeshop Skunk appear many times bigger than it really is. Moreover, all of the panels are decorated with semi-transparent strips in various shades of green. A reference to the botanical substances on sale. 'It's as if you are surrounded by a fairytale glade of abstract cannabis stalks,' is how Mentjens explains the experience.

Completion Date: 2014

为了遵守当地严格的规则和条例,委托方希望将两个完全独立的咖啡馆融合在一个建筑里。尽管允许开设内部通道,每个咖啡馆也都要有各自的出入口。由于场地呈不甚方便的三角形,再加上这些要求的限制,就产生了一个不同寻常的平面图。

外卖店"Skunk"将道路一侧的街景尽收眼底,两侧通道通向位于建筑后部的"Relax"。这两个通道占用了大量的空间,即便两个店面共享卫生间、服务区和安全亭等设施,通道还是割裂了整个零售面积。

"我们将通道改造成壮观的体验空间,将缺点变成优势。"设计师表示。其中的秘密就是面板上的镜面。设计师将镜面朝向外墙,透明一面朝向内侧。最终的效果使得Skunk咖啡馆看起来比实际大得多。此外,所有面板都饰有不同深浅的半透明绿色丝带,彰显店内所售商品为植物产品的理念。"看起来,就好像深处植物茎干的抽象世界一样",设计师如是解释道。

建成时间:2014年

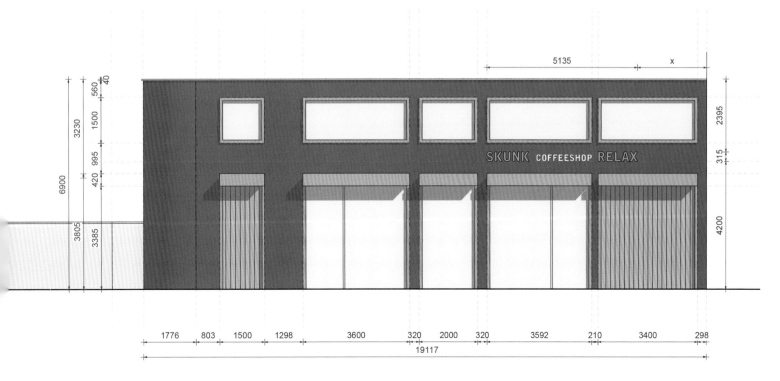

AAP (ASSOCIATED ARCHITECTS PARTNERSHIP)

DALLAH

Kuwait City, Kuwait

杜拉咖啡厅

AAP 设计事务所 / 科威特，科威特城

Dallah is a traditional coffee pot used to brew and serve 'qahwat dallah'; an Arabic type of coffee specifically known to the Gulf or the Arabian Peninsula region.

An asymmetrical arched form was designed to further accentuate the warm central 'heart' of the shop and at the same time to define the exterior vitrine and entry experience. The asymmetrical white arch's springing line defines two very different situations on either end; a smaller arch that gives more of a human scale to the outdoor space just before entering the shop flanks one side of the springing line while a larger more fluid arch expands into the interior of the shop giving a sense of intimacy and openness to the indoors.

Completion Date: 2015   Area: 45m²   Photographer: João Morgado

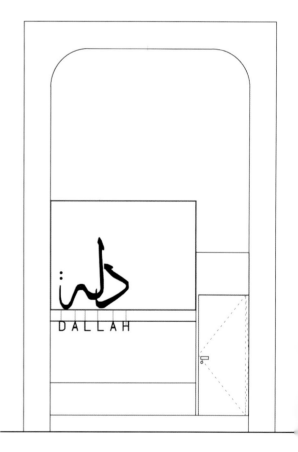

杜拉是用于煮制和奉上阿拉伯咖啡"qahwat dallah"的传统咖啡壶，在海湾和阿拉伯半岛地区尤为常见。

不对称拱形设计进一步强调了店铺核心部分温馨的"心脏"，同时限定了外部玻璃橱窗的范围，设定顾客的进门体验。这个不对称白色拱形结构的线条两端分别代表两个截然不同的情况：较小的拱形让室外空间更加人性化，较大的拱形以更为流畅的形式延伸至店内，同时赋予室内空间亲密性和开放性的特点。

建成时间：2015年　面积：45平方米　摄影：若昂·莫尔加多

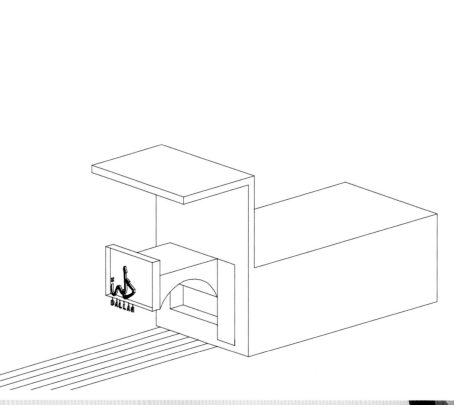

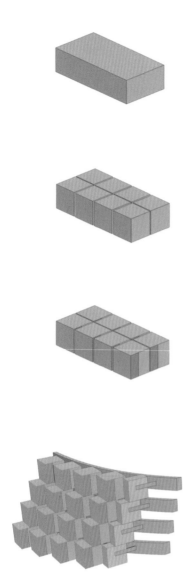

HOOBA DESIGN GROUP

ESPRISS CAFE

Tehran, Iran

Espriss 咖啡馆

Hooba 设计集团 / 伊朗，德黑兰

The aim of the project was to renovate a gift shop and change it to a cafe. Considering the small size of the project and its location, the main idea was inspired by the urban context to transform the traditional elements into an architectural interior space. In designing the spatial diagram, the materiality concept is based on an integrated geometry continuing from outside to inside. The neighbouring building, Iranian handicrafts organization with brick facade, was the inspiration to use the same material for the cafe. Concerning the small size of the project, a brick with 5×10×20 dimensions sliced into eight smaller pieces of 5×5×5 centimeters which one side of the bricks glazed in turquoise blue colour. The terracotta bricks are also hygienic as they covered with antibacterial layer. The 3D diagram created an integrated structure for all the features required to work out the issues of the existing structure which also introduce morphology of brick an light as all the bricks are positioned based on the 3D diagram. The lightening of the project is formed by the lamps provided in the gaps in between the joints of the bricks inspired by Iranian traditional architecture which translated into a modern language of design.

Designer: Hooman Balazadeh Project Manager: Elham Seyfiazad Design Team: Niloofar Altaha, Noushin Atrvash Completion Date: Summer 2014 Area: 28m² Photographer: Parham Taghi-of

本案的目标是翻新一家礼品店，将其改造成一间咖啡馆。考虑到项目规模较小及其所在位置，主要设计思想着眼将传统元素转换成具有建筑感的室内空间。空间设计图的制作过程是按照从室外到室内的顺序进行的。相邻的伊朗手工砖墙建筑激发了咖啡馆的设计灵感。设计师将5厘米×10厘米×20厘米的砖块分割成5厘米×5厘米×5厘米的8个较小单元，砖块的一侧是蓝绿色玻璃。赤褐色的砖上也添加了抗菌涂层，更加卫生。三维设计图中的集成结构包含了原有建筑亟待解决的问题，也加入了砖石结构与光线的形态关系。项目中的照明设计分布在砖石连接处的缝隙中，以现代设计语言呈现伊朗传统建筑元素。

设计师：胡曼·巴拉扎德 项目经理：伊尔哈姆·赛飞亚扎德 设计团队：妮露法·阿尔塔哈，诺申·阿特瓦什 建成时间：2014年夏 面积：28平方米 摄影：帕勒姆·塔基夫

BROLLY DESIGN

FLIPBOARD CAFE

Melbourne, Australia

Flipboard 咖啡馆

布罗利设计公司 / 澳大利亚，墨尔本

Flipboard Cafe, calved from a lost site in the city, is a tiny multi-level nook that serves fine space with a side of excellent coffee and healthy food. The cafe is nestled in the intersection of an emergency exit from Bennetts Lane Jazz Club below, the thoroughfare to Brolly Studios behind, and a two decade old unused shop-front window.

The space had once been a transient uncentered passage but has now become its own destination whilst still maintaining its other activities. In the renovation of the front facade the designers found spaces that had been lost to previous weak and unimaginative occupants who had plasterboarded it shut. They opened it, and it is now one of the few eddies that interrupt the flow of foot traffic on La Trobe street as you can inhabit the facade in cosy cubbies.

Design Team: Martin Heide, Megg Evans   Area: 25m²   Photography: Tanja Milbourne, TM Photo

Flipboard 咖啡馆所在的位置是城市中一处被人遗忘的空间，是一个供应优质咖啡和健康食品的多层角落。

这里曾经是一条短暂的通道，如今虽然还保留着其他活动功能，但本身已经成为一个别致的空间。店面翻新过程中，设计师发现了前人用石膏板封闭起来的隐蔽空间并将其打通，打造成沿街的趣味景观。

设计团队：马丁·海德，麦格·埃文斯   面积：25平方米   摄影：塔尼娅·米尔本，TM 图片公司

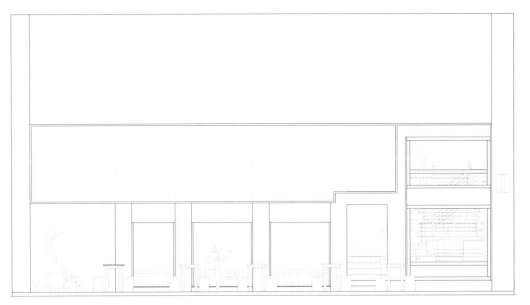

ESTUDIO VITALE

JULIETA, PAN & CAFÉ

Castelló de la Plana, Spain

Julieta, Pan & Café 咖啡馆

比塔莱建筑事务所 / 西班牙，卡斯特利翁 - 德拉普拉纳

Create a genuine business, where the protagonists are core products: bread and coffee. Enhance an honest and simple image of respect for tradition, craftsmanship and dedication of a well-crafted product. Under these assumptions Estudio Vitale designed the brand and the retail unit.

A vintage style graffiti on the facade reinforces the traditional and authentic business image. Large windows reinforce the friendly and transparent business spirit and provide visibility within the local, in true European style. From the street the business is perceived like an honest and simple image of respect for tradition, craftsmanship and dedication of a well-crafted product. The project greatly respected the original bricks facade of the building. Brand values are transmitted by austere and simple materials at the facade. The combination of white and black reinforces the corporate identity of the local. The facade is complemented by labeling brand inspired by the paper doilies used for traditional cakes. It creates an honest and homely image from the street to invite people to come and have a coffee.

Completion Date: September 2013 Square-foot Shop: 80m²
Square-foot Lab: 20m² Photographer: Estudio Vitale

打造以面包和咖啡为主角的企业。面对传统、工艺和奉献，强化一个真诚、简单、充满敬意的形象。设计师以这些概念为基础，进行了品牌形象和零售空间设计。

店面外墙上的复古风格涂鸦强化了品牌传统而又真诚的商业形象。大扇的窗户增强友好、透明的商业精神，以欧式风格明显的方法增加与周围环境的互动。设计师在项目中充分保留了原有的砖石外墙，并通过质朴简约的外墙材料表达品牌价值观。企业形象在黑白配色中得到升华。

建成时间：2013年9月　面积：（店面）80平方米，（实验室）20平方米　摄影：比塔莱建筑事务所

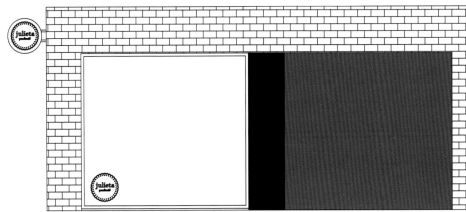

TALLER KEN

'MAJADAS ONCE'

Guatemala City, Guatemala

"Majadas Once" 咖啡厅

Taller Ken 设计公司 / 危地马拉，危地马拉城

'Saúl Bistro, Majadas Once' mines patterns, materials and textures and collects them to make a fresh tropical atmosphere. One of the biggest design decisions is that the café is open-air, with no glass, to connect the project to the pedestrian walkways. The exterior consists only of custom operable folding shutters comprised of steel panels inter-woven with dowels. It 'skins' the existing commercial center's facade to create a greater impression of a unified whole when seen from all 3 free sides.

Completion Date: 2014 Area: 460m² Photographer: Andres Asturias

"Majadas Once" 咖啡厅使用了图案、材料和不同的材质，打造了一个让人耳目一新的热带环境。设计过程中的最大挑战在于这个完全开放的咖啡厅里没有连接人行通道的玻璃结构。店面外墙仅由定制的可折叠百叶窗构成，依附在原有商业中心外墙上，从三个方向观察都是统一的协调外观。

建成时间：2014 年 面积：460 平方米 摄影：安德烈斯·阿斯图里亚斯

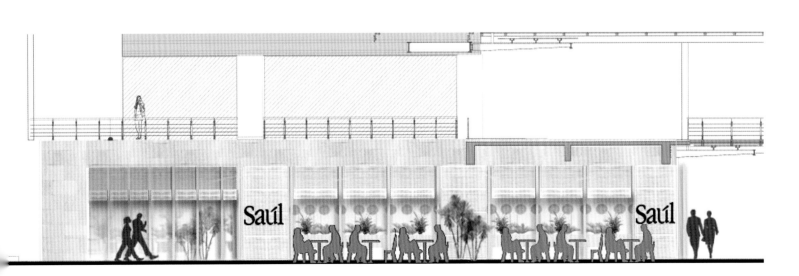

STUDIO DOTCOF

TERRACE CAFÉ – FOOOO

Chengdu, China

临街露台咖啡店——孚乐里

多棵设计 / 中国，成都

Early in his career, Frank Lloyd Wright experimented the relations between interior and street, on both Cheney house and Roby house. He built a wide terrace between private living room and street. Raise the terrace slightly above street level and protect it with a low wall, which you can see over if you sit near it, but which prevents people on the street from looking into the living room.

Studio DOTCOF use this pattern to build a big terrace connecting the interior and exterior, raise the indoor and outdoor floors up on a same level, and place planters surrounding the terrace. The only division inserted in between is a full-length sliding glass door. When the glass door is fully open, interior and exterior space are merged completely. The only solid partition is a wood box mass on the big terrace, which includes a kitchen, a dishwashing room and a small intimate dining area.

Area: 110m² Completion Date: 2015.2 Photographer: Mir Kim, Hao Cheng

赖特曾在他从事建筑的初期，在切尼住宅和罗比住宅等设计中，对室内空间与街道之间的关系进行了实验：他在私密的起居室和公共的街道之间，建造一个很宽的露台，使露台略高于街道，并以低墙保护它，如果你坐在露台上，你可以从墙的上方望见街景，但墙却阻挡了街上行人的视线，使他们看不见室内空间。

Studio DOTCOF 多棵设计应用这一模式，搭建了一个连接室内外的大露台，将咖啡店室内外抬升到同一个标高，室外部分的四周以植物环绕。室内外之间几乎没有设置任何实体的阻隔，只是加入了一道通长的、可完全打开的玻璃推拉门。当玻璃推拉门打开时，室内空间与室外露台就完全融为一体。唯一的墙体是大露台上一个木盒子体量，它包含了备餐厨房、洗碗间和一个相对私密的就餐区。

面积：110平方米　施工时间：2015年2月　摄影师：金美玲，程颢

YOSHIHIRO YAMAMOTO
CAFÉ FRANZ KAFKA

Nara city, Japan

弗朗茨·卡夫卡咖啡馆

山本义弘 / 日本，奈良

This is the conversion of a traditional Japanese residence into a book cafe in Nara World Helitage Site. Tributed to The Shosoin Repository which kept many ancient treasures from overseas for a long time, Yoshihiro Yamamoto designed interiors as an antique trunk with old timberframe and dark shining fabric.

The façade design followed Machiya style (traditioal Japanese townhouse) under regulations of cityscape in Nara World Helitage site. Main material is Kawara (Japanese roof tiles), Shikkui (Japanese Stucco), and exposed Japanese cedar columns and beams. Only a white cube facing the entrance is modern element. It unites tradition to modern, fuses East and West.

Area: 61.31m² Completion Date: Junly 2015 Materials: Walls: Plywood, polyurethane paint, tile; External walls: Shikkui (Japanese stucco)

本案中，奈良世界文化遗产地的一间日本传统住宅被改造为书店咖啡馆。设计师向珍藏了许多古代海外宝藏的正仓院致敬，利用木质结构和深色反光面料打造出古董行李箱一般的室内空间。

外墙设计遵循"町屋"风格（传统日式排屋）以及奈良世界文化遗产地的相关规定。项目使用的主要材料是日式瓦片、日式灰泥和柳杉立柱。正对入口的白色立方体是唯一的现代元素。它将传统与现代、东方与西方联系在一起。

建成时间：2015年7月　面积：61.31平方米　材料：内墙：胶合板、聚氨酯涂料、瓷砖；外墙：日式白色抹灰

I LIKE DESIGN STUDIO

STARBUCKS FOOD VILLA

Bangkok, Thailand

星巴克食品屋咖啡厅

I Like 设计工作室 / 泰国，曼谷

Starbucks Food Villa is located at Food Villa Market Ratchapruek, Bangkok. The building is a 2-storey steel structure featuring drive-thru, bar seating on the ground floor and meeting room, toilets on the 2nd floor. As it is one of Food Villa Projects, the design concept of Starbucks flagship store needs to be outstanding but in harmony with the neighbourhood. The main concept of Food Villa project is a Food Mall. The Market represented 'farmhouses', developed from farmhouses elevation pattern and retail. Design concept is inspired by an important contrast between the luxurious of Starbucks brand and the simplicity of farmer house. Neat and modern gable roof on the ground floor represented the farmer house style as other building in the project. Transparent glass box is placed on the top to look clearly from outside and to glow at night from inside to show users' lifestyle.

星巴克食品屋咖啡厅位于曼谷的食品屋市场的两层建筑内，包含汽车餐厅、一楼有吧台座椅，二楼是会议室和卫生间。作为食品屋的设计项目之一，这间旗舰店的设计理念需要既出众，又与环境协调。食品屋的主要设计理念是打造一个食品商场。市场代表的"农舍"使用农舍外墙的图案，零售网点代表的则是"农民之家"。

该设计理念灵感来自星巴克品牌的奢华形象与农民之家的简洁形象之间的鲜明对比。一楼采用的整洁现代的人字形屋顶设计代表了农民之家的风格。结构顶部的透明玻璃结构增加项目的通透性，夜晚则会在灯光照耀下熠熠生辉。

Completion Date: October 2015 Photographer: Soopakorn Srisakul

建成时间：2015 年 10 月   摄影：Soopakorn Srisakul

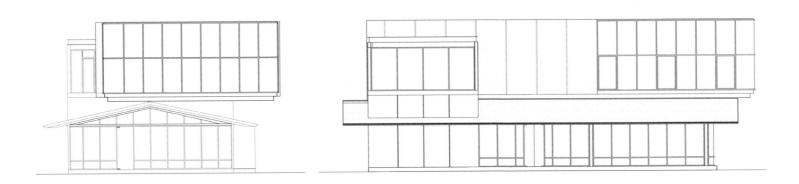

STUDIO RAMOPRIMO

TRAITOR ZHOU'S

Beijing, China

Traitor Zhou's 综合餐厅

RAMOPRIMO 设计工作室 / 中国，北京

Traitor Zhou's is the first of a chain of little deli corners in Beijing. The challenge for the project was to insert a simple but yet elaborated functional program consisting in market, bakery, café and whisky bar into an existing room of only 16 sqm.

On the façade, the specific requirement from the owner to create a young urban feeling for the new shop has been also answered by using triangular elements in the primary colours of red yellow and blue. The reference of the traditional Japanese female dress 'yukata' has been used to provide the space of a new dress and visual identity which could embody the different kinds of food offered, location and personal background of the owner, which is a mix of different western and eastern influences. The entrance door is the only opening to the outside. The design of the façade will enlarge this element by adding two iron 'sleeves' on the sides, which can work as external counter for a quick coffee or a glass of Japanese whiskey. Traitor Zhou's logos are made of paper and have been installed like a poster on the façade.

Completion Date: 2015 Design Team: Marcella Campa, Stefano Avesani Area: approx 18m² Photographer: Marcella Campa - RAMOPRIMO Finishing Materials: Iron plates, plywood, concrete grass block pavers

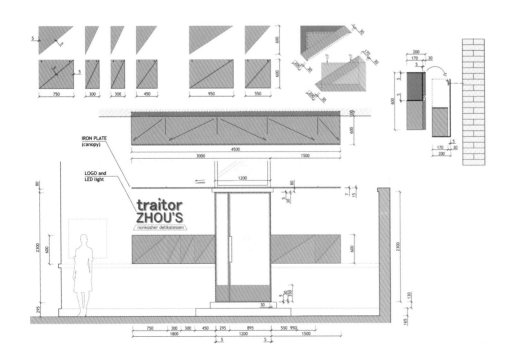

这间Traitor Zhou's综合餐厅是连锁迷你餐厅在北京的第一家分店。设计中存在的挑战是如何在仅有16平方米大小的空间里容纳形式简单但功能丰富的店面内容，包括商店、面包店、咖啡馆和酒吧。

设计师通过在外墙上使用红黄蓝三色的三角形元素满足了店主打造年轻化都市感店面的愿望。使用传统的日本女性服饰元素代表店内经营的不同种类的食品，店主的个人背景，实现了东西方文化的融合。大门是店面与外界的唯一通道。外墙上在大门两侧增加的两个铁"袖"使门变得更大，新结构也可以当做外部吧台使用。店面标识由纸制成，以海报的形式出现在外墙上。

建成时间：2015年　设计团队：玛塞拉·坎帕，斯特凡诺·伊乌萨尼　面积：约18平方米　摄影：玛塞拉·坎帕——RAMOPRIMO设计工作室　材料：铁板、胶合板、混凝土草块铺面

LYCS ARCHITECTURE

UNDERLINE

Hangzhou, China

Underline 咖啡店

零壹城市建筑事务所 / 中国，杭州

The project is located in the Eastern Software Park which lies in the center of Hangzhou. Although surrounded by high-rent office buildings, there is no public space for communication or meeting in this area. To meet the companies' needs of working, meeting and communicating, Underline could be changed immediately from a single-function cafe into a cafe + meeting room or cafe + office room through free combinations of its interior spaces.

Instead of over-decorated design, the designers created a peaceful atmosphere for the cafe in a modern and concise way. The facade of the cafe is made of all-glass windows with chessboard-like black frames. The concrete walls above the facade are set with the LED logo of 'underline'. The design is quite simple but ingeniously fits into the surrounding environment. In order to bring out a vivid effect, the designers decorated the large-scale black wall with relaxing and humorous painting.

项目位于杭州市中心的东部软件园内，园区内外汇集了大量的创业公司，周边租金不菲却几乎没有适合洽谈的场所。为了满足这些公司对于办公、会议和交流的需求，Underline通过空间的自由组合，可瞬间将单一功能的咖啡厅变身为Cafe+会议室或Cafe+办公室。

为了使咖啡厅拥有更好的办公氛围，项目摒弃了传统咖啡厅花哨的设计，采用了简洁现代的设计手法。咖啡厅外墙采用了全玻璃幕墙与棋盘般的黑色窗框设计，幕墙上方的混凝土墙面安置了"underline"字样的LED logo，简明的设计使项目和谐的融入于周围环境中。

为了让外立面显得更加活跃，设计师在幕墙两旁的黑色墙面上设计了轻松的墙画，将咖啡厅的功能通过幽默的方式进行直观的表达。

Project Date: 2014 (Design) 2015 (Construction+Completion) Design Team: Ruan Hao, He Yulou, Jin Cheng, Fu Li Area: 240m² Image Copyrights: LYCS Architecture

项目周期：2014（设计）至 2015（施工并完工） 设计团队：阮昊、何昱楼、靳成、傅立 项目面积：240平方米 图片版权：零壹城市建筑事务所

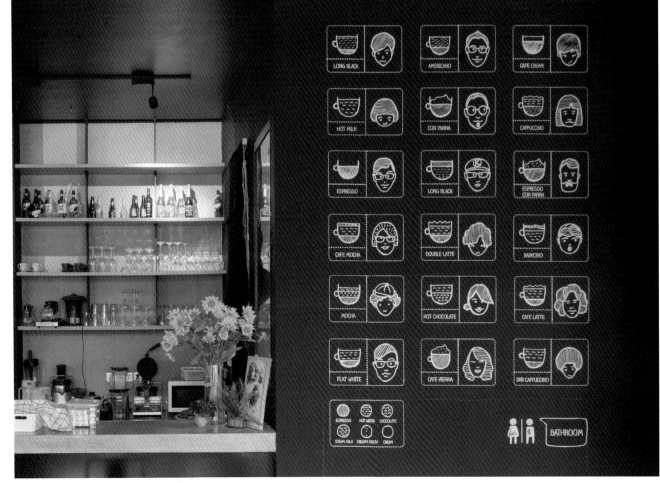

HITZIG MILITELLO ARQUITECTOS

TOSTADO CAFE CLUB

Buenos Aires, Argentina

Tostado 咖啡俱乐部

希齐西·米利泰洛建筑师事务所 / 阿根廷，布宜诺斯艾利斯

The complete facade needed to be recycled with wooden windows painted in black to keep the monochrome style. The designers wanted also to keep the old style lighting. At the window there are drawings of old crockery, as well as cups, toast machines that allude the concept of the space. The facades that face to the avenue show a take away window that can be opened or closed. The products that are exposed were prepared and specially designed by a chef. The menu is gourmet sandwiches and differents types of coffee.

Design Team: Magdalena Molinari, Juan Carpinello Construction Company: Estudio Cores
Completion Date: 2015 Size: 180m² Photography: Federico Kulekdjian

本案中的正面外墙以及刷成黑色的木窗需要回收，以便保持店面的黑白风格。设计师还希望保留旧式的照明装置。橱窗位置绘有老式陶器、茶杯、烤面包机，彰显空间的个性理念。朝向街道的外墙上设置了一扇外卖窗口，可以开合。店内的产品都经过厨师的精心设计和烹制，不仅有美味三明治，也可以找到不同种类的咖啡饮品。

设计团队：马格达莱纳·莫利纳里，胡安·卡佩内罗 建筑公司：科雷斯建筑公司 建成时间：2015年 面积：180平方米 摄影：费德里科·库勒德江

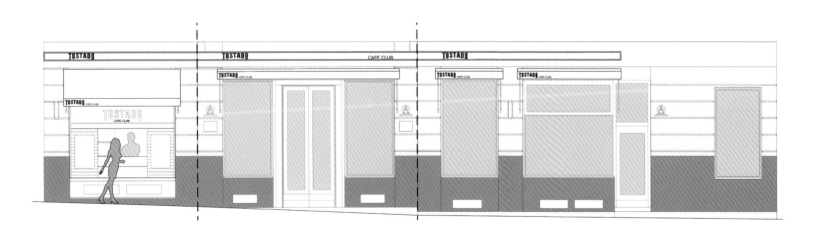

169 • 咖啡厅、酒吧

VAWDREY HOUSE

CULLENDERS

London, UK

Cullenders 酒吧

Vawdrey House 设计公司 / 英国，伦敦

The Vawdrey House worked closely with Cullenders to transform their dream of opening a successful full-scale, eat-in cafe, deli and wine bar, into reality. Cullenders occupies a prominent high street location in this traditional market town. The three-storey 1930s building sits in a run of shops varying from medieval to Victorian to 1970s.

Hand-painted signage, lantern lights and an eye-catching striped awning gives the deli a strong street presence, but with a sense of establishment, which is perfect for its location. The large shop front window provides clear views into the interior, and to the owner behind the counter always ready with a smile. The shop is welcoming and flexible – transforming effortlessly from the morning hustle and bustle of a coffee bar and general store to a relaxed space for afternoon tea and later to a lively evening venue for drinks and food.

Completion Date: 2014   Photographer: Siobhan Doran Photography

本案中 Vawdrey House 设计公司与 Cullenders 酒吧密切合作，共同打造一个成功的集堂食咖啡馆、熟食店和红酒店于一体的餐饮场所。Cullenders 酒吧坐落在这个传统集镇的高街上。这间始建于 20 世纪 30 年代的三层建筑位于一处建于中世纪、维多利亚时期到 20 世纪 70 年代的商铺之间。手绘的店铺招牌、灯饰和引人注目的条纹遮阳篷赋予这间店铺强烈的街景存在感，但又具有一种历史感，与周围环境相符。

宽大的店面橱窗方便人们了解室内环境，与柜台后总是面带微笑的店主人相识。这是一个温馨、友好又颇具灵活性的店面设计——早间熙熙攘攘的咖啡店和杂货铺可以毫不费力地转换为适合享受下午茶的悠闲空间，而到了晚间就变身成为热闹酒吧。

建成时间：2014 年　摄影：西沃恩·多兰摄影公司

WANG TAO

TSINGTAO 1903 BAR

Qingdao, China

TSINGTAO1903 酒吧

王涛 / 中国，青岛

This is the starting point of Community Pub Plan of Tsingtao Beer Company and it is also the very first store. An experiential chain store of beer culture is born and it is low cost, fast to replicate, of unified identity and most importantly it is also environmentally friendly, as almost all of the materials are recycled or reused.

The doors of the pub can be opened in two different ways. During hot season, the door can be widely open so that people will have a better flow in-and-out. During cold seasons, people can enter through the small door for better insulation. The entire glass window cannot be opened and it is for pedestrian to see inside the bar more clearly. The counter around operation area makes meeting new friends more easily. The challenge is to create a youthful space and at the same time it also needs to represent the historical values of Tsingtao beer brand. Under limited project budget and relatively small construction site, the creative idea is to design the bar, both internal and external, in a simple and geometry stereo segmentation style. A large entrance door is also used to weaken the division between the outside space and inside space.

Area: 69m² Photographer: Wang Tao

这是青岛啤酒公司"社区酒吧计划"的起点，也是第一家店面。一个可以低造价、快速复制、具有统一识别性的连锁的啤酒文化体验店由此诞生。最重要的是它也很环保，几乎所有的材料都采用回收或可再利用。

这个酒吧的大门可以根据不同季节或活动的要求采用两种开启方式。夏季的时候全部开启，使室内和室外的间隔性减弱，人们可以更好的沟通和流动；冬季天气寒冷，人们用小门进出，更加保温。窗户为不可开启的整玻璃，路过的行人可以更方便的看到酒吧内的人们，围绕操作区的吧台为与新朋友交谈创造了便利。正对街面的酒瓶墙也采用回收再利用的酒瓶。

项目面积：69平方米　摄影：王涛

STUDIO RAMOPRIMO

BUONABOCCA - ITALIAN WINEBAR

Beijing, China

啵拿博卡意式红酒店

STUDIO RAMOPRIMO 工作室 / 中国，北京

Beginning with store name and logo design, the client, an Italian-Chinese couple, wishes to use 'mouth' and elements related with food and erotics as the leading theme. They also required yellow resin floor. Therefore the store exterior adopts a consistent colour scheme of black and yellow with the interior. The element of mouth is applied as a graphic element in custom-made interior and exterior wallpaper. The logo outside hangs over a row of seating created with the repeating colours of black, yellow and white. They provide resting place for the customers as much as passers-by, and at the same time adding volume and functionality to the exterior design.

在红酒吧的名称和标志设计上，我们的客户，一对中意夫妇，向我们提出希望以"嘴"同时与食物和情色相关联的元素为主题，并要求使用黄色树脂地板。店铺外立面与室内要相互呼应，所以外立面的配色也是由黑色和黄色完成。"嘴"同时也成为了一个平面装饰元素，被印制在店内以及外立面的订制墙纸上。外立面 logo 下面摆放了一排座椅，由黑色、黄色、白色组成，显然，是为了搭配这间红酒店的主题而定制的。这一排座椅的作用，不仅可以让食客们舒适的度过等位时间，也可以供过路人临时休息，更是为店铺的外立面增加了立体感、实用性，使这家店铺更聚人气。

Completion Date: 2016　Design Team: arch.Marcella Campa, arch.Stefano Avesani
Area:60m²　Photographer: RAMOPRIMO

建成时间：2016年　设计团队：玛塞拉·坎帕，斯特凡诺·埃伏萨尼
面积：60平方米　摄影：STUDIO RAMOPRIMO 工作室

ADAM WIERCINSKI ARCHITEKT

ŹRÓDŁO.BAR

Poznan, Poland

Źródło 酒吧
亚当·威尔辛斯基建筑事务所 / 波兰，波兹南

Źródło.bar is a new whiskey bar on the map of Poznan. It's located in the basement of a tenement house on Taczaka street - one of the most popular streets with restaurants and cafes in the city center. For the last few years this place was empty hence it was in a very bad condition. After cleaning it up, the original brick walls and wooden-brick floors showed up. Main idea was to preserve original look of the place with addition of the new, independent structure. Together with the investor, the designers decided that each element of this interior, including light and furniture, will be designed individually. They wanted to reduce the use of ready-made products. The designers placed three steel structures inside the bar — these are boxes holding all of the elements like seatings and lighting. None of the elements, except surfacemounted electrical installation, has been screwed into original walls. The designers used subtle, graphite colours with orange accents which do not compete with materials and intense brick colour.

Area: 45m² Photographer: Przemyslaw Turlej

Źródło 酒吧是波兹南一家新开的威士忌酒吧。它位于塔扎卡街一间公寓的地下室内，这里是市中心最热闹的街道之一，布满了餐厅和咖啡厅。过去几年间，项目所在的位置一直闲置，因此条件很差。清理之后，原有的砖墙和木地板都显现了出来。项目中主要的设计理念是在保护原始建筑的基础上，增加新的独立结构。设计师与委托方共同决定独立设计包括照明和家具在内的所有室内元素，以减少对预先做好的产品的使用。设计师在酒吧内部安置了三个钢结构——这些箱式结构将容纳座椅和照明等所有室内元素。除了安装在表面的电气装置，其他元素都不是固定在原有墙壁上的。设计师巧妙地使用石墨与橙色，与其他材料和砖面浓烈的颜色相映成趣。

面积：45平方米　摄影：波尔兹米斯洛·特尔基

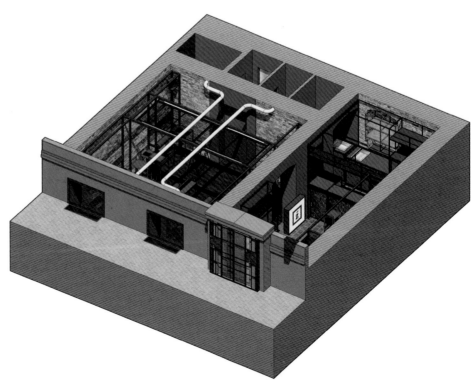

REIICHI IKEDA DESIGN

8B DOLCE FOSHAN

Foshan, China

佛山 8b DOLCE 甜品店

池田励一设计工作室 / 中国，佛山

A popular dessert brand from Korea, 8b DOLCE recently opened its new branch in Foshan, Guangdong. It is located in an old building with grey brick exterior and kept the existing wooden beam, in the renovated Foshan Lingnan New World Shopping Mall. A giant poster hanging outside reads '佛山特产冇得弹', meaning 'extraordinary Foshan specialty', which is a typical Lingnan expression itself.

The store takes up no more than 62m², yet the reputation of 8b DOLCE means the products will never let you down.

Completion Date: January, 2016 Area: 61.97m² Photographer: Yoshiro Masuda

韩国甜品店 8b DOLCE 已经在网络上活跃已久，最近它的新店开到了黄飞鸿的老家广东佛山。这家店开在由旧建筑改造而成的佛山岭南新天地，门面是一座青砖结构的老房子，并保留了房顶的木梁。而店外的墙壁上挂着一幅巨大海报，写着"佛山特产冇得弹"，即"佛山特产不得了"的意思，广东岭南风味十足。

门店本身不大，只有不到 62 平方米。而 8b DOLCE 名声在外，蛋糕的味道也不用再赘述。佛山岭南新天地如今是个颇为好逛的地方，特色餐厅和咖啡店不少。如果下次去广州或者香港游玩，或许可以考虑下绕道到佛山一趟。

建成时间：2016 年 1 月 面积：61.97 平方米 摄影：益田喜郎

KENGO KUMA AND ASSOCIATES

SUNNY HILLS

Tokyo, Japan

Sunny Hills 烘培店

隈研吾建筑都市设计事务所 / 日本，东京

This shop, specialized in selling pineapple cake (popular sweet in Taiwan), is in the shape of a bamboo basket. It is built on a joint system called 'Jiigoku-Gumi', traditional method used in Japanese wooden architecture (often observed in Shoji: vertical and cross pieces in the same width are entwined in each other to form a muntin grid). Normally the two pieces intersect in two dimensions, but here they are combined in 30 degrees in 3 dimensions (or in cubic), which came into a structure like a cloud. With this idea, the section size of each wood piece was reduced to as thin as 60mm×60mm.

As the building is located in middle of the residential area in Aoyama, the design team wants to give some soft and subtle atmosphere to it, which is completely different from a concrete box. The designer team expects that the street and the architecture could be in good chemistry.

Date: 2013.12 Area: 297m² Photographer: DAICI ANO

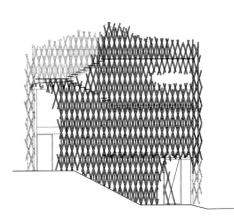 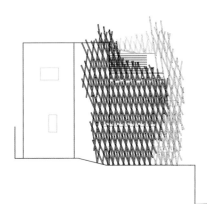 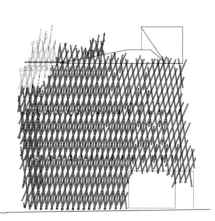

本案中的店铺是一家菠萝蛋糕（台湾流行甜品）专营店，外观呈现竹篮的形状。采用的连接系统名为"Jiigoku-Gumi"，是日本木建筑中的一种传统技术（宽度相同的窄木条在垂直和水平两个方向交叉打结，形成一个木条网格）。通常是两块木条在二维平面上相互交叉，但在本案中的结合方式呈30度角，采用三维立体编织成的云朵一般的外观。出于这样的技术考虑，每个木条的尺寸减为60毫米×60毫米。

由于建筑位于青山居民区中央，设计师希望打造出一种完全不同于混凝土方块，而是柔和、微妙的气氛。在设计师的眼中，街道和建筑之间会发生奇妙的化学反应。

竣工日期：2013年12月　面积：297平方米　摄影：DAICI ANO

NAN ARCHITECTS

BAKERY MAISQUEPAN

Galicia, Spain

Maisquepan 烘焙坊

NAN 建筑师事务所 / 西班牙，加利西亚

The project stems from the idea of the owners to create a multifunctional facility, which seeks to enhance their bakery business, incorporating the cuisine of Portuguese and Brazilian origins, offering suitable for a short stay tasting their products and coffee areas. Besides typical bread sale where rotation would be very high.

The owners had some concerns about the design of the facility, and they did not want a conventional bakery and sought the establishment itself give them a good projection as brand image.

The idea was to create a facade that was very striking, where the operation of the entire office can be seen from the street. This is a very open place where the outer and inner communication was very fluid, making the stay in the premises a pleasant experience and the street people feeling encouraged to enter.

A choice of materials was to combine steel and wood. The facade steel marked where interruptions façade emphasize the aspects designers want to highlight. In this case an exhibition of wines, a bar where people look to the street and inside the counter, where you see people doing their own jobs of this type

of establishment. Thus the door shown in the closed part searching with this contradiction highlights the other elements.

Both outside and inside, steel is used as a search for the material forward aseptic feeling.

Lighting also has sought a careful design. The designers have sought shelter metaphorically to evoke a large mass and this they have done with molds of the same material that mimic those of a loaf of bread, and for the work area, these molds are an organic element where the 'mass' seems to fall to the ground. They have also opted for signage lighting and concealed lighting to enhance the wood paneling in, and lighting in all the cases.

Design Team: Wenceslao López, Vicente Pillado, Alberto Reiriz Area: 94m² Photographer: Iván Casal Nieto

本项目始于店主打造一个多功能设施的想法，他们希望通过设计提升烘焙坊的业务，综合源自葡萄牙和巴西的美食，为顾客提供一个适宜短期停留，品尝店内烘焙产品和咖啡的区域。

店主对设计方案有些担忧，因为他们并不想要一个传统风格的烘焙坊，而是希望借助这次设计打造一个优质的品牌形象。设计师的想法是打造一个吸人眼球的外墙，让人们在街上就可以看到整个操作间。这将是一个非常开阔的空间，无论外部还是内部流通都非常顺畅，顾客在这里停留的过程中将拥有一段愉悦的体验，路人也会被吸引进店。

材料选择方面，设计师选择结合使用钢铁和木材。人们在店外可以看到店内的红酒展示以及吧台的情况，与周围反差强烈的大门显得其他元素更为出挑。

设计团队：温西斯劳·洛佩兹，韦森特·皮拉多，阿尔贝托·雷利兹
面积：94平方米　摄影：伊凡·卡萨尔·涅托

MOVEDESIGN

BOULANGER KAITI

Tamagawa, Fukuoka, Japan

卡提面包店

MOVEDESIGN 设计公司 / 日本，福冈，多摩川

This is a bakery shop renovated from an old Japanese house in Fukuoka, Japan. The area is a residential district close to the downtown. The existing house built of wood has impressive old wood walls and tile-roofing. As for neighbouring houses, one could see this house has a history and exist as a symbol of this town. The design team started to plan not to appeal the facade as a shop renewing the whole image of this building but regarded to match with surroundings as important and let the shop fit this residential district, arranged facade design to integrate new section into old building, and furthermore, configured so that people can recognize there is a bakery shop here.

They replaced only one surface of wall on south side with the shop section with all white appearance not to be imbalanced in existing characteristics. Those wide lining windows attract interest of customers from the outside, from the inside, let the sun light moderately in and limit the view to outside so that customers can shop comfortably. The owner's philosophy, 'to offer delicious quality bread everyday', means he does not permit compromise on everything, selecting ingredients or how to make bread. His passion and his personal character percolate through this place and create 'Kaiti's view of the world' here.

Designer: Mikio Sakamoto   Area: 138.05m²   Photogragher: Yousuke Harigane /Techni Staff

这间烘焙坊是由日本福冈一间老旧的日式住宅翻新而成,位置处于靠近市中心的住宅区域。原有住宅由木头建造,有漂亮的木墙和瓷砖屋顶。与周围的建筑相比,项目所在的建筑有一定的历史,称得上是城镇上的一个符号。设计师开始着手设计时,考虑的不是在店铺设计中让外墙更加耀眼,而是通过融入周围环境,通过外墙设计将新的元素带入旧的建筑,从而进一步确定面包店的存在感。

设计师只将店面南侧的一面墙刷成白色,为了不与原有环境特征相冲突。宽大的窗户一方面吸引顾客的兴趣,另一方面让日光有控制地照射室内,让顾客可以舒适地购物。

店主的理念是,"每天提供美味的高质量面包"。他不允许任何方面的妥协,无论是原料挑选还是制作工艺。他的热情和个性渗透进了店铺的方方面面,打造出了"卡提的世界观"。

设计师:坂本干雄　面积:138.05平方米　摄影:Yousuke Harigane/ 技术人员

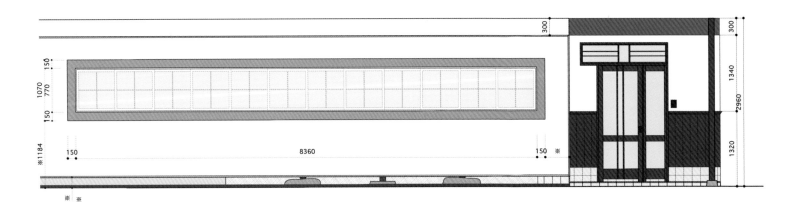

MORENO: MASEY

GAIL'S BAKERY, WARDOUR STREET

London, UK

沃德街盖尔烘焙坊
moreno: masey 设计公司 / 英国，伦敦

In the heart of the vibrant Soho district of London and on a prominent corner site, the design approach was to transform the historic listed property into a showcase for the Client's exceptional bakery products. With functionality paramount and mindful of the high footfall in this area, the interior was refined to maximize the display of the produce while ensuring an open, well lit and spacious interior.

Existing rough textured brickwork was white washed, while bright clean and smooth corian was selected to form the feature bar. Areas of the bakery were then deliberately separated using a warm palette of unique materials. A social space at the rear of the property was fitted with a large table with a delicately designed iron base and lit by an abundance of bell-shaped ceiling lights to draw the clientele into the property. Side bars continued the material with the use of matt iron and the different areas are seamlessly blended together with subtle changes in finish and bespoke joinery.

Photographer: Richard Lewisohn

项目位于充满活力的伦敦 soho 区中心地带，一个著名的转角位置。设计师希望通过改造历史保护建筑，打造一个展示委托人优质烘焙产品的窗口。秉着功能至上的理念，考虑到这个区域较高的客流量，室内设计以实现产品展示的最大化为目标，同时保证开放、明亮和宽敞的空间。

原有的粗糙纹理砖砌结构经过洗刷磨白，选用明亮干净且光滑的可丽耐大理石打造特色吧台。烘焙操作区域有意使用暖色材料区分。店面后侧的公共空间安装了带有精致钢铁基座的大桌子，多个钟形的天花板灯具投下适宜的光线，吸引顾客进店体验。侧面吧台沿用了亚光钢材，在不同区域进行微小调整和定制处理，无缝地融为一体。

摄影：理查德·拉维生

ARQUITETURA DAVID GUERRA

LA GALETTE

Brazil

La Galette 餐厅与烘焙坊

大卫·圭塔建筑事务所 / 巴西

The initial concept of the project was to reinstate the status of a buffet and patisserie with 25 years of market, while providing it with a new identity. La Galette's project makes an allusion to affection, singularity and refinement, seeking to portray the identity of old European patisseries with a modern touch. The golden cabinets were made with various door designs and mirrored decorations, giving greater spotlight to the exposed products. The golden tint is at the same time seductive and welcoming, while the pink brings a touch of irreverence to the space. As a contraposition to the colours, blue, grey and black appear on the cement tiles, bringing modernity while still making reference to an antique conception. The expositor on the window shop, designed by the architect as two linear pieces on a sequence of lathed supports, also reveal the identity of La Galette, bringing an irreverent form of presentation. A handmade scissor window secures the window shop.

Area: 70m² Photographer: Jomar Braganca

项目的初始理念是为有着25年历史的自助餐和法式糕点店重新焕发活力，同时为其打造一个新的品牌形象。这是一场对情感、独特和精致的追求，试图用现代手法描绘出一个历史悠久的欧洲糕点师形象。金色柜子上出现了多种门的设计和镜面装饰，增强了产品表面的光泽。金色同时也具有诱惑力和友善度，粉色则为空间增添一丝叛逆。与之相对的蓝色、灰色和黑色出现在水泥砖上，引入现代元素，也融汇复古概念。橱窗的展示设计是由一系列板条支撑的两个流线结构，同样展现着店铺充满个性的独特的品牌形象。手工制作的剪刀式窗口保证橱窗的牢固。

面积：70平方米 摄影：宙马·布拉干萨

ESTÚDIO JACARANDÁ ARQUITETURA + DESIGN DE VAREJO

OJUARA TAPIOCARIA

Sorocaba, Brazil

Ojuara Tapiocaria 木薯食品店

Jacarandá 事务所（arquitetura + design de varejo）/ 巴西，索罗卡巴

Estúdio Jacarandá has created the concept for the new retail brand Ojuara Tapiocaria, dedicated to the sale of tapioca, a typical Brazilian food with indigenous origins heavily consumed in the northern and northeast regions of the country. The architectural, branding and visual merchandizing projects for the store are inspired by the history of tapioca, which is shown by the functional elements in this concept, such as the wall cabinets that reference the façades of the northeastern outback and the wall menu adorned with woodcut images. Featured are also objects handmade by indigenous communities and the counter lined with certified pine wood with serigraphy that mimics the hand-painted ornaments found on trucks. The colour pallet, composed by green, orange and magenta colours, reinforces the Brazilian northeastern aesthetics.

Designer: Luciana Carvalho, Renato Diniz   Area: 12m²   Photographer: CEZAR KIRIZAWA

Jacarandá 事务所为全新的零售品牌 Ojuara Tapiocaria 食品店打造了店面理念。Ojuara Tapiocaria 是一家木薯专营店。木薯是在巴西北部和东北地区十分热门的一种巴西经典食品。店面的建筑设计、品牌形象和视觉营销设计都从木薯的历史中汲取灵感，并通过功能性元素得到展示，例如令人联想到东北内陆地区房屋外墙的壁柜，以及由木刻图案装饰的墙壁菜单。土著社区手工制作的物件和模仿卡车上手绘饰品的绢印松木工艺品。店内绿色、橙色和品红色的配色组合使得巴西东北地区风格进一步得到强化。

设计师：卢恰娜·卡瓦略，雷纳托·迪尼斯   面积：12平方米   摄影：塞萨尔·桐泽

PARTY/SPACE/DESIGN

SHUGAA ROOM FOR DESSERT

SUKHUMVIT 61, Bangkok, Thailand

Shugaa 甜品馆

party/space/design 设计公司 / 泰国，曼谷

The concept of the design of Shugaa had been researched from the base of sugar which its form is sugar molecules and crystals. Seeing from the outside through the glass wall, there is polygon installation hanging around the front that is inspired from sugar crystals. Besides, wood material also had been used in the design together with mint green colour to make it feel warm and earthy. Designer team have added a dash of modern and luxury by using rose gold colour element and marble counter bar.

Completion Date: January 2016   Area: 100m²   Photographer: F Sections

Shugaa "糖" 甜品馆设计理念的相关研究从糖的分子和晶体等基础开始。透过玻璃墙壁从室外观察，可以看到多边形悬挂装置，灵感来自糖的结晶。此外薄荷色木质材料的使用让室内空间变得温馨朴实。设计团队利用玫瑰金元素和大理石吧台为设计增添几分现代奢华的气息。

建成时间：2016年1月   面积：100平方米   摄影：F Sections 公司

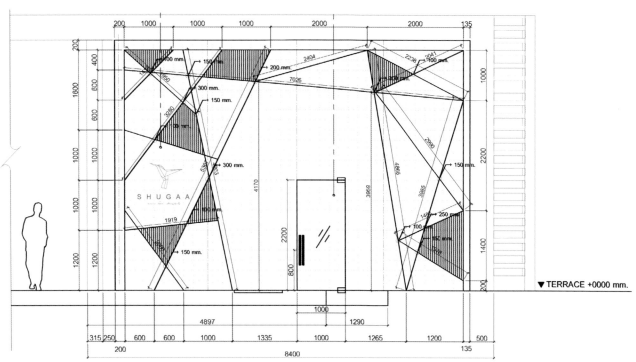

PROCESS5 DESIGN

UCHIYA BAKE SHOP HANATEN BIRD'S HOUSE

Osaka, Japan

内山烘焙屋

PROCESS5 DESIGN 设计公司 / 日本，大阪

Under the shop concept/name based on 'loved by many' and 'jibun no uchiya (means "my home" in Kansai dialect)', the store No. 2 of this egg tart shop is now open. They opened the store in front of Hanaten station that has good access to the downtown area comprised with mixture of new apartments and old rows of houses. This area is also home to residents of all ages ranging anywhere from families to senior citizens. By taking advantage of the location that gives illusion of as if the store is standing in the middle of the road when seeing it from the station, the design team wanted this new 'Uchiya' to be a kind of store that can give impact and enjoyment to people living in the city, while preserving the cost effectiveness and remnant of former clock shop.

Through the method of cutting 2x4 white wood material being sold at regular hardware stores to a fixed length, and by attaching it to the existing appearance at a fixed angle, and as the final step combing them together, the design team were able to come up with a kind of expression that can stand out in the city as well as bringing warmth to the city. The curious view of the store that can be seen from the crack between the trees along with the fragrant aroma, are sure to lure city's passer-by into the store.

Completion Date: October 2014   Design Team: Ikuma Yoshizawa, Noriaki Takaeda, Taichi Fujiwara   Area: 18.53m²
Photographer: Stirling Elmendorf Photography

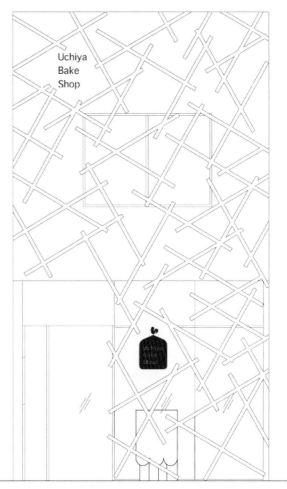

秉承"广受喜爱"和"我的家"（日本关西方言）的店铺理念和店名，这个蛋挞品牌的首家分店正式与顾客见面。店铺选址于 Hanaten 车站站前，到市中心交通便利，那里不仅有新建公寓也有老式排屋。这一区域的居民分布于各个年龄段。设计师充分利用店铺在车站位置观察似乎位于道路中央的地理优势，希望将这间新"内山"店面打造成能够产生影响力并为居住在城市中的人们带来愉悦的店铺，同时保留原钟表铺子留下的实用元素。将普通五金店出售的 2x4 尺寸白色木料切割成固定长度，以一定角度安装到原有结构，最后合为一体。设计师得以创造出一种在城市景观中脱颖而出的表达方式，同时为城区带来友好温馨的氛围。通过树木间隙观察店铺，伴随芬芳香气所看到的奇异景象想必会吸引众多路人进店。

建成时间：2014年10月　设计团队：吉泽奈央，武田德昭，藤原太一　面积：18.53平方米　摄影：斯特灵·埃尔门多夫摄影公司

ANONYM

BAAN KANOM CHAN

Bangkok, Thailand

Baan Kanom Chan 面包店

Anonym 设计事务所 / 泰国，曼谷

The design team used French windows for the south façade, expanding the view of nearby forest and promoting natural ventilation. In front of the building, a double balcony is added to function as the entrance. This is a space of multi-functions, where the client and the staff can carry out all sorts of activities freely.

The application of a variety of materials is another feature of the design, such as glass, black steel framework and white plaster walls. The contrast between the building and its surroundings makes it stand out in the neighbourhood.

Design Team: Phongphat Ueasangkhomset, Suteenart Chantarajiraporn, Parnduangjai Roojnawate   Area: 600m²   Photographer: Chaovarith Poonphol

建筑师在南面设计了落地窗，扩大了面对附近树林的视野角度，同时也促进了自然通风。在房子的前面，建筑师设计了一个双阳台，创建了主进出口。这是一个多功能空间，业主和办公室工作人员可以在此开展任何活动。

该房子的另一个突出的特点是在设计上使用了多种材料，比如玻璃，黑色钢框架和白色的灰泥墙。房子与周围环境之间形成的反差让该建筑在社区中独树一帜，成为了一栋在该地区引人注目的建筑。

设计团队：冯帕·乌桑克欧姆赛，苏提纳特·占他拉吉拉普恩，班当贾伊·卢杰纳瓦特   面积：600平方米   摄影：乔瓦利斯·鹏波

BAREA+PARTERS

POLLOS PLANES STORE

Valencia, Spain

波略斯·普拉内斯食品店

Barea+Parters 设计公司 / 西班牙，瓦伦西亚

The shop front restyling comes from the need to expand the business to more demanding markets and the project's scope is to transform the image of the current company stores into a fresh, clean, contemporary style that expresses quality and proximity. So B+P creates a facade that includes the corporate identity, and the interior and architectural design of each element; a facade that is aware of the brand's tradition and long experience in the food sector.

In order to improve the clients shopping experience the B+P team has created a series of corporate elements that are part of a strategy that takes advantage of the queuing times to bring to light the company's expertise. Some of these elements are created with the added scope of increasing the target market, reaching the smallest customers. Thanks to wisely selected illumination, and carefully designed counters (heroes of the store), the new improved environment created in the store transmits quality to the product.

Regarding the corporate image and although the implementation of the signage has dramatically changed in relation to the former format, the new logos have been designed in a way that continuity between the old and new image is easily readable, making it simple to identify the restyling of the new company and being the change in the logos unnoticed for the non-trained eye.

Completion date: 2014

本案的店面翻新项目源自业务扩展的需要,以适应愈发苛刻的市场。项目范畴包括将现有的公司形象转变成清新、简洁且现代的新形象。Barea+Parters 设计公司因而打造了一个包含企业形象、室内设计和建筑设计元素的外墙方案;一个体现品牌传统以及品牌在食品领域长期经验的外部设计。

为了提升顾客的消费体验,设计团队还设计了一系列企业元素,充分利用顾客排队等候的时间,作为宣传策略的一部分。其中一些元素的设计还包含了增加市场份额,争取每一位顾客的思想。由于有了巧妙选择的照明装置和精心设计的吧台,升级后的新环境中产品看起来质感更佳。

企业形象方面,尽管店面标识的安装方式十分显眼和夸张,还是可以在新标识设计中感受到新旧形象的联系和传承。

建成时间:2014 年

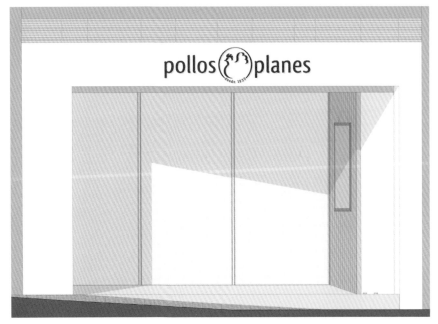

JGV

HERSHEY'S LAS VEGAS

Las Vegas, USA

拉斯维加斯好时巧克力世界

JGV 设计公司 / 美国，拉斯维加斯

Located at the New York-New York Hotel & Casino, Hershey's Chocolate World becomes a focal point of New York-New York. The facade features an 18-feet diameter illuminated Reese's Peanut Butter Cup. Visitors will also be greeted by a 60-foot wide and 24-feet tall high resolution digital monitor marquee. The array of LED illuminated 15 feet tall Kisses dynamically 'suspended' above the consumer visually punctuates the architecture of the storefront, and connects the key elements of the facade, including the super-over scaled six-storey high Hershey Bar. The store has a series of multiple access points to the Las Vegas Boulevard strip, including the key entrance through a gigantic Hershey Milk Chocolate Bar, plus two additional casino interior entrances. There are a variety of iconic fixture elements that draw their inspiration from the products in the Hershey's brand portfolio, ranging from the oversized Hershey's Kisses-shaped merchandise fixtures, to the Reese's feature displays and walk through elements; leveraging the highly identifiable and proprietary colours and materials unique to each of the brands.

Completion Date: June 2014   Area: 13,000 square feet   Photographer: Laszlo Regos Photography

好时巧克力世界位于纽约-纽约酒店＆赌场内，是一处引人注目的场所。店面外部是直径 18 英尺的灯光锐滋花生酱杯。游客还会看到 60 英尺宽，24 英尺高的高清数字显示器。15 英尺高的 LED 好时之吻悬挂在高处，为整个建筑结构增加视觉亮点，并将外墙的超大、六层楼高的好时巧克力棒等主要元素联系在一起。店内有多个出入口通往拉斯维加斯大道，其中包括穿过巨大好时牛奶巧克力棒的主要出入口，以及两个赌场内部入口。此外，店内还有多种多样的标志性元素，灵感均来自好时品牌的经典产品，比如超大的好时之吻造型装置，锐滋形象的展示和穿行设计；均采用各个品牌辨识度较高的形象和专有配色及材质。

建成时间：2014 年 6 月   面积：13,000 平方英尺   摄影：拉斯洛·瑞格思摄影

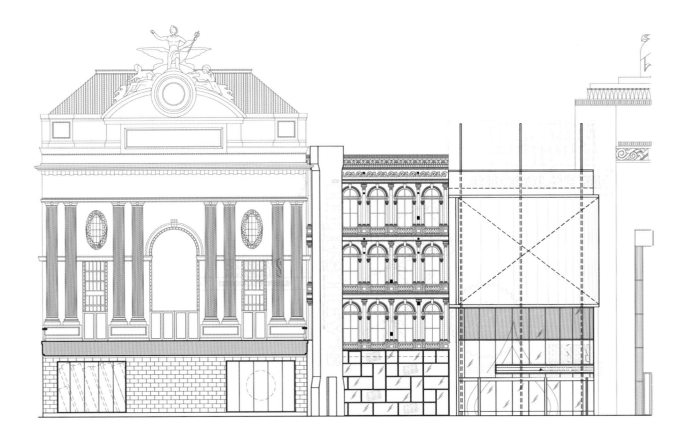
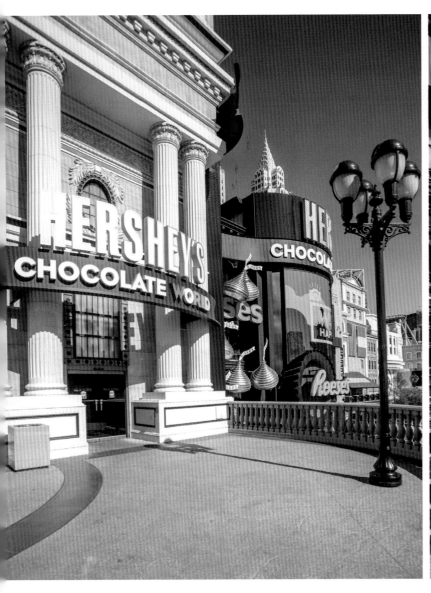

NOT A NUMBER ARCHITECTS

JUICE BAR CABIN

Bucharest, Romania

果汁酒吧小屋

Not a Number 建筑师事务所 / 罗马尼亚，布加勒斯特

The juice bar occupies a small site in the center of Bucharest. Its irregular layout follows the outline of a derelict shed that was previously standing on the plot working as a Turkish café. The space was conceived as a cabin in the city, a safe haven from the bustling streets or the cold afternoons. The cabin's exterior walls are uniformly coated in standing-seam black sheet metal. While dark and monolithic during the day, when the sun comes down two large openings animate the front façade with the activity inside. Creating the impression of an inviting house through form, as well as materiality was of a high priority. The dark shell smoothly unveils the warm interior through specific frames of its attractive core. At the same time, the contrasting element with the take away window emerges from the metal skin and brings out the indoors atmosphere, increasing curiosity towards the wooden heart.

果汁酒吧小屋所在的位置是布加勒斯特中心的一处小空间。这块不规则场地曾经是一家土耳其咖啡厅。这里被视为城市中的一间小屋，一个远离熙熙攘攘的街道和寒冷下午的安全避风港。小屋外墙统一采用了黑色金属板材覆面，白天的时候外部看起来深沉单一，天黑以后两扇大窗展示室内活动，使得店面外观也跟着活跃起来。形式和材料的选择上都力争打造温馨的氛围。深色的外墙下展露的是温馨的室内空间。外卖窗口的对比元素出现在金属外墙上，将室内氛围加以延伸，吸引人们进店探索店内木质装饰的温馨世界。

Completion Date: 2014   Designer: Ermis Adamantidis, Dominiki Dadatsi   Area: 120m²
Photographer: Cosmin Dragomir

建成时间：2014 年   设计师：艾米斯·安达曼提迪斯，多米尼基·达达兹   面积：120 平方米   摄影：科斯明·德拉戈米尔

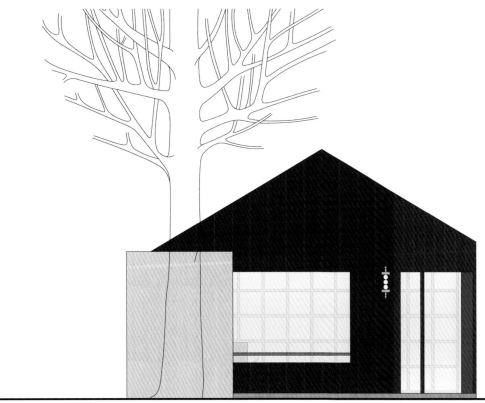

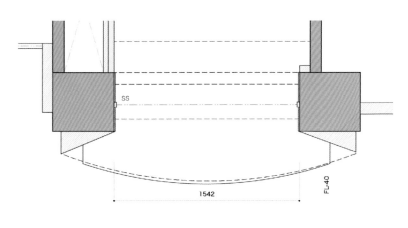
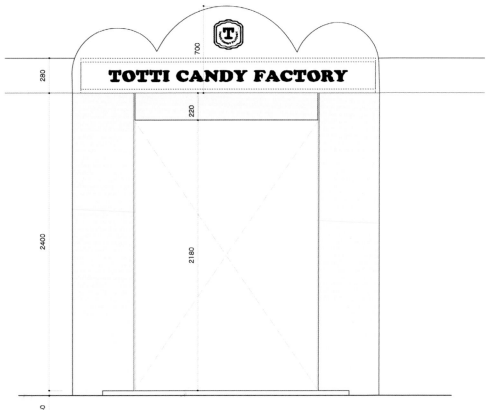
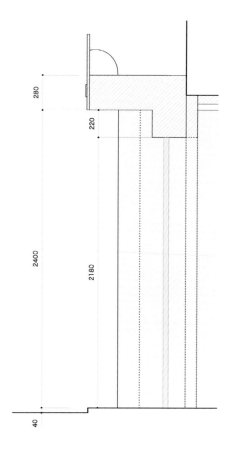

MOUSETRAP

TOTTI CANDY FACTORY

Osaka-shi, Japan

托蒂糖果工厂

mousetrap 设计公司 / 日本，大阪

'Totti Candy Factory' is the shop which sells sweets and candies from all over the world. Mainly they sell candy, chocolate and candy floss.

Mousetrap planned their facade to make an impression of cuteness and amusement to the customer. Pink and white colours help the shop concept and make their brand strongly. And the facade sign will give an impression of candy floss and it helps that people become curious about the shop easily.

"托蒂糖果工厂"是一间出售全球生产的各式糖果的商店，主要经营硬糖、巧克力和棉花糖。

设计师计划为顾客打造一面呈现可爱有趣风格的店面外墙。粉色和白色深化店面理念，让品牌形象更为鲜明。外部标识让人联想起棉花糖，吸引人们对店铺产生好奇心。

Completion Date: February 2015   Designer: Hiroyuki Saiki   Area: 34.7m²   Photographer: Hiroyuki Saiki

建成时间：2015年2月   设计师：佐伯洋之   面积：34.7平方米

ARNAU ESTUDI D'ARQUITECTURA

FERRER XOCOLATA

Gerona, Spain

费雷尔巧克力

阿尔瑙建筑事务所 / 西班牙，赫罗纳

The designers had to renovate an old building in order to create a space entirely devoted to chocolate. That was the sweet dream of Jordi, the chocolate maker. The facade used chocolate tile moulds used in chocolate making. The designer team relied on the staircase and its visual omnipresence as well as on the different passing gaps that link vertically the three lower levels.

Everything has been specially designed in this project: from the section of the building to the chairs in the tasting area or the chocolate tiles. It is a global and absorbing project, which turned out to be very addictive.

Completion Date: April 2015   Designer: Arnau Vergés Tejero   Photographer: Marc Torra_fragments.cat

本案中，设计师需要将旧建筑翻新，打造一个完完全全的巧克力世界。这是巧克力制造商乔迪的甜蜜梦想。外墙使用的装饰元素灵感来自制作巧克力的模具。设计师利用楼梯的视觉效果和各式通道，将较低的三层空间联系起来。

项目中使用的所有材料都为特别定制：从品尝区的座椅到巧克力瓦片。这是一项国际化的项目，对任何人来说也是极具魅力的。

建成时间：2015 年 4 月   设计师：阿尔瑙·维杰斯·德杰罗   摄影：Marc Torra_fragments.cat 摄影

TACET CREATIONS

FRESCO GELATERIA

New York, USA

弗雷斯科冰激凌店
tacet 设计工作室 / 美国，纽约

The design was for a concept store of a new brand in NYC called Fresco Gelateria. The owner intention was to create a Greek / Hamptons hybrid allowing emotional access for the American public. This was achieved by using surfaces reminiscent of the Greek islands such as the stucco and white colours & the gently cream colours of the distressed woods to remind of the New England shacks and beach structures. The lighting was very important as the design team created a feeling of light filtering from above and around through slots between the wood planks. Ceramic lights by Lightexture with a white glazed finish added to the ice cream aesthetic and white corian was used for the counters.

The main design elements were the lowered floor so that the store is 'looked into' from above and turning of the ice cream display so as to enhance the mystery of what is on display and make the clients come in rather than view directly from the street. This is what had contributed to the success of the relaxed ice cream brand rather that the direct presentation that gives away the story too fast.

Designer: Nicholas Karytinos (design) Alina Ainza (lighting)
Photographer: Domagoj Blazevic

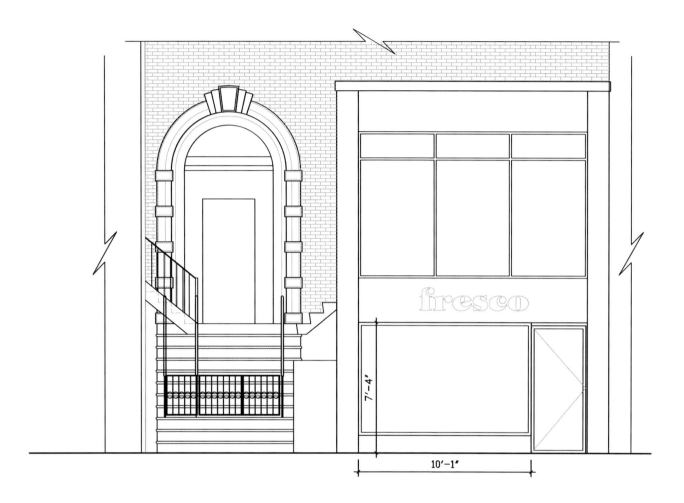

本案是为纽约一家名为弗雷斯科冰激凌店概念店所作的设计。店主希望打造一个希腊/汉普顿混合风格的店面,可以与美国公众建立情感联系。为了实现这一目标,设计师在店面设计中使用了灰泥和白色等让人联想到希腊岛屿的元素,并通过做旧木材的乳白色体现新英格兰的棚屋和沙滩。设计师十分重视照明设计,利用木板条之间的缝隙打造过滤效果。Lightexture 生产的陶瓷灯具外层是白色玻璃表面,呼应冰激凌的店面主题。柜台则使用了白色可丽耐大理石。

下降地板称得上是项目中的主要设计元素,"俯视"的视觉效果使得冰激凌展示更具神秘感,吸引消费者进店探索。这个设计理念使得店铺的经营十分成功。

设计师:尼古拉斯·凯瑞蒂诺斯(设计),爱丽娜·艾因扎　摄影:多麦格·布拉泽维奇

EXTERNAL REFERENCE ARCHITECTS WITH MANUEL ATECA

HAM ON WHEELS

Barcelona, Spain

轮子上的火腿

External Reference 建筑师事务所，曼纽尔·阿特卡 / 西班牙，巴塞罗那

Ham on wheels is a premium Catalan fast food within a tiny urban space full of colour, aromas and flavours. The project decomposes the basic 'ingredients' of the concept reassembling them into a new spatial experience. The ham becomes an iconic element that when repeated, produces a luminous landscape ceiling. The hanging Hams differ in colour, graphics and display different messages.

餐馆外立面采用了工业美学的设计手法，将涂鸦作为装饰效果的一部分。同时还将美味的食材元素融入到室内、外的设计中来，火腿是本店的专售产品，也是本店的独有装饰元素，透过玻璃门就可以看到，它通过不同的颜色来表现，并添加展示信息。也为就餐者带来了轻松好玩的用餐体验。

Completion Date: November 2015 Built Area: 40m² Photographer: Lorenzo Patuzzo

建成时间：2015年11月 面积：40平方米 摄影：洛伦佐·帕图佐

WATT INTERNATIONAL

HÄAGEN-DAZS

Hackescher Markt, Berlin, Germany

哈根达斯旗舰店

瓦特国际／德国，柏林

The Häagen-Dazs prototype shop was conceived as a space of supple sophistication, where the community of passionate ice cream lovers could come together, bask in the glow of the brand, and share in spectacular culinary creations. Its storefront, a protected facade in the heart of Berlin's Hackescher Markt, carries a black, halo lit wordmark. The simplicity of the identification respects the austerity of the historic architecture, while the colours, lighting and finishes of the interior sparkle. The effect is a window into the brand's restrained sensuality, inviting passers-by to enter and indulge.

Completion Date: 2014   Photography: Fiona Dunsmore, Tobias Wille

这间哈根达斯旗舰店是一处精巧而复杂的空间设计。狂热的冰激凌爱好者会聚集在这里，沐浴在品牌的光辉中，分享让人惊叹的美食产品。坐落在市场中心的店面外墙上是黑色的品牌标识。简洁的设计元素充分体现对历史建筑的尊重，同时又在色彩、照明和墙面材质上透露室内元素。

建成时间：2014 年　摄影：菲奥娜·邓斯莫尔，托拜尼斯·威利

LABVERT
DEPARTMENT STORE WELS

Wels, Austria

韦尔斯百货商店

LABVERT 设计公司 / 奥地利，韦尔斯

This is a project of proportions and dimensions adjusted to neighbouring buildings. When, in the late 19th century, the first department stores opened in Europe's big cities, this was the first time that retailers sought to attract customers with glazed storefronts, large shop windows and passageways. Technological innovation made it possible to open up closed facades and let the urban street life passing by get a glimpse of the glittering world of consumer goods. For the design of the storefront in Wels, LABVERT developed classical department store architecture further by affording not only a view of the world of consumer goods, but also of the sky above.

Eight generously dimensioned windows dominate the square-grid facade. With their frames protruding from the wall at different angles, facing different directions, they break up the rigid alignment pattern of the facade. The buildings around and across the street are reflected in the window panes. Townscape and sky, street life and goods display all flow into one another, the clear geometry of the building fits in with the ensemble of the neighbouring facades as well as with the paving-stone pattern in the street in front of the building. But that's not all yet.

Designer: Stephan Vary Completion Date: January 2014 Photographer: Lisa Rastl

19世纪末，欧洲大城市中的第一家百货商店建造时比例和尺寸参照邻近建筑设计，如今是零售商们第一次试图用玻璃店面、巨大的橱窗和宽敞的通道吸引顾客。技术创新使得打开封闭的墙壁，让城市街道上经过的人们浏览橱窗里琳琅满目的消费品成为可能。韦尔斯百货商店的店面设计项目中，LABVERT 设计公司在更深层次上开发了经典的百货商店式建筑形式，不仅展示种类繁多的商品，还将建筑顶上的天空纳入设计范畴。

方形网格外墙的主要结构是 8 个巨大的窗户。窗框以不同角度从墙上突出，朝向不同方向，打破了僵化、对齐的外墙格局。周围以及街道对面的建筑也映在窗户玻璃上。城市景观和天空、街头生活以及商品展示相映成趣，建筑清晰的几何形态与相邻建筑的外墙造型以及百货商场正对街路上的铺面图案协调相融。

设计师：斯蒂芬·瓦里 建成时间：2014 年 1 月 摄影：丽莎·拉斯特尔

NORDIC BROS. DESIGN COMMUNITY / YONG-HWAN SHIN

NORTH GATE SALON

Seoul, South Korea

北门沙龙

北欧兄弟设计工作室 / 申永焕（音译）/ 韩国，首尔

The North Gate Salon is a select shop in Yeonhee-dong. This space puts together business and everyday conversation with the neighbourhood. The client found out 'Kafe Nordic' through her own research. For that, she became interested in Nordic Bros's works and visited those places. Finally, she asked 'Nordic Bros. Design Community' for 'The North Gate Salon project'. Most of all, the designer was fascinated by the tiny space at the first meeting with the North Gate Salon, and started the project hoping to solve the limitations or inconveniences of the space. The North Gate Salon was first designed on pleasant imagination and boundary of unreality, so that it might draw humour of that region. However, the project couldn't overcome of the limit which is building itself. After a short period of blank, the designers modified the concept; gable, arch, and horizontality, with the second design concept, then he was able to complete the North Gate project.

Due to the characteristics of the building exterior where only the minimal renovation is allowed, it concentrated on the windows and doors and focused on harmony and balance with the interior adjusting the location, size, and form while maintaining the shape of the previous structure.

Completion Date: Febuary 2015 Construction: Nordic Bros. Design Community / Sung-Won Park Area: First floor 13m^2, second floor 7m^2, eranda 8m^2 Photographer: Nordic Bros. Design Community

北门沙龙是位于首尔延禧洞的一家精品店。设计在环境中融合了商业文化与日常交流。委托方在搜索调查时发现了"Kafe Nordic"咖啡厅并由此对北欧兄弟设计工作室的作品产生了兴趣，逐个进行了走访。最后邀请北欧兄弟设计工作室为自己设计打造一个"北门沙龙"项目。最关键的是，设计师在第一次查看北门沙龙项目所在场地时，被如此狭小的空间吸引，希望通过设计解决场地面临的限制和不便。北门沙龙的设计理念以愉悦想象和虚幻界限为基础，会为周围环境带来不同寻常的幽默感。然而项目本身难以克服的问题是建筑本身。遭遇了短期的瓶颈之后，设计师对设计理念进行修改：山墙、拱门和水平方向，项目最终采用第二个设计概念完成。

由于建筑外部只允许进行小幅改造，设计师将注意力转向门窗，通过在保持原有结构基础上调整室内装饰的位置、大小和形式达到和谐平衡的目的。

建成时间：2015年2月 施工公司：北欧兄弟设计工作室/朴成媛（音译）
面积：一楼13平方米，二楼7平方米，阳台8平方米 摄影：北欧兄弟设计工作室

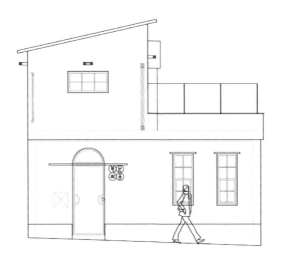
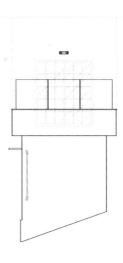

IO STUDIO, ING. ARCH LUKA KRIŽEK, ING. RADEK BLÁHA

VODAFONE DATA SHOP OSTRAVA

Ostrava, Czech Republic

斯特拉瓦沃达丰数据商店

IO 设计工作室，卢卡·克里泽建筑公司，拉杰克·布拉哈设计公司 / 捷克，斯特拉瓦

This concept of facade elevates and enhances the products displayed, creates a volumetric difference from the planes and lines and points bringing greater visibility to the designer pieces. Facade lines relate to the lines forming the inner planes. The designers used new materials and technology that is completely beyond the classical concept of stores. Well-known corporate logo 'drop in the circular pattern' was developed from a 2D graphic display in the 3D irregular helix. Logo dynamically surrounds and defines the layout of space to different zones.

本案中，外墙的设计理念提升了产品展示的质感，在点线面的基础上打造出体量感极强的空间，为产品带来更大的可视性。外墙的线条与内部平面采用的线条相近。项目中使用了超出传统店面范畴的新材料和新技术。意为"圆形陷阱"的知名企业标识从二维的平面展示升级为三维不规则螺旋形式。店面标识以动态的形式丰富了不同区域的空间布局。

General contractor: DCH-Sincolor a.s.　Photographer: Alexander Dobrovodský

总承包商：DCH-Sincolor a.s. 公司　摄影：亚历山大·德布洛沃斯基

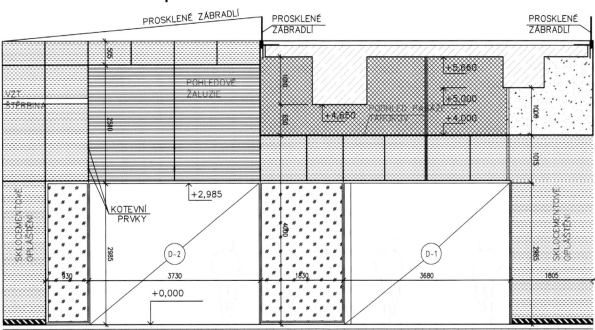

MASQUESPACIO

DOCTOR MANZANA

Valencia, Spain

曼萨纳医生

Masquespacio 设计公司 / 西班牙，瓦伦西亚

The logotype starts from the principal axe of the company 'the touchscreen' and his reflection that creates an angle of 54 degrees. That angle ends being part of the whole communication and his defragmented into different applications that create an infinity of forms for the graphic and interior design. Ana Milena Hernández Palacios, creative director of Masquespacio: 'Talking about the colours as we started from a company name allied with a doctor we wanted to create a concept based on a hospital, however as we didn't want to create a conventional design, we discarded this option, but maintaining blue and green colours as a reference to the first word in the company's brand name.' Looking at the store everything starts from the striking facade that incorporates the same angles and colours like for the graphic identity. The blue and green colours like a reference to the doctor, the salmon colour for the fashionistas and the purple for the freaks.

Designer: Ana Milena Hernández Palacios   Photographer: David Rodríguez

本案中的企业商标由公司的斧子形象和与之构成54度角的倒影组成。这一概念得以通过多种不同形式应用于平面和室内设计中。Masquespacio设计公司的创意总监安娜·米莱娜·赫尔南德斯·帕拉西奥斯表示，"由于公司名称中出现了'医生'，起初我们想要创造一个与医院意象相关的店面形象。然而这一定不会是一个中规中矩的设计，所以我们放弃了这个方案，只保留蓝色和绿色代表医生的意象。"可以看出，店内所有的设计都与引人注目的墙面设计一样采用了相同的角度和配色方案。蓝色和绿色让人联想到医生的形象，粉橙色前卫时尚，紫色小众个性。

设计师：安娜·米莱娜·埃尔南德斯·帕拉西奥斯　摄影：大卫·罗德里格斯

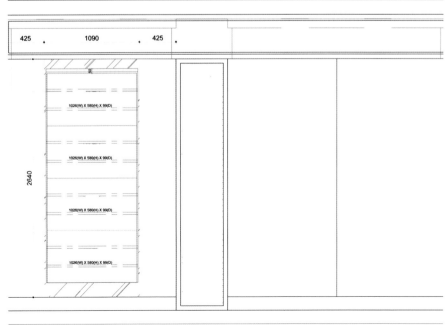

DAAA HAUS

KLIKK CONCEPT STORE

Zejtun, Malta

Klikk 概念店
DAAA HAUS 设计公司 / 马耳他，泽尔顿

The concept behind the new Klikk store was to introduce colour and exciting elements to the commonly mundane computer store. Pop Retro Culture was the main inspiration, which pumped the establishment with vibrant colour, comical imagery, dynamic forms, Pop Art. The playful approach to the new concept was aimed at bringing a more vibrant atmosphere and enhancing the customer experience to computers.

When it came to tackling the designs for the facade of Klikk two very important matters were kept in mind, the first was to create delivery truck access to the storage area without hindering the aesthetics of the design and the second was to give a hint of the pop-art brand identity in a clean and minimal design. The first design solution came about by cladding the whole of the facade with black aluminum horizontal bars, including the garage door, to create a flush hidden door within the structure. This in turn, gave a perfectly uniform large, black canvas to place the bright and bold logo signage. The contrast between the dark linear background and colourful logo was created to radiate an effective eye catching brand which pops, as the store is situated in front of a very busy main road. To the right of the façade, large glass apertures are used to give a clean open view of all the playful pop inspired designs happening internally.

Designers: DAAA & LOGIX CREATIVE TEAM Completion Date: December 2014 Area: 500m² Budget: €250,000

将色彩和令人兴奋的元素引入普通电脑商店是 Klikk 概念店背后的设计理念。流行复古文化是设计的主要灵感来源，为项目注入了充满活力的色彩，趣味的图像，动感的形态和流行文化。实施新概念采用趣味手法是为了营造更加活跃的气氛，提升顾客使用电脑过程中的体验。

设计团队在外墙的设计处理上考虑了两个非常重要的因素，一是在不影响设计美观的前提下给送货卡车留出便捷的通道，二是在简约的设计中适当体现品牌形象的流行艺术元素。第一个解决方法是在包括车库门在内的整个外墙上使用水平黑色铝条覆面，形成隐藏门。这也成为了明亮店面标识的大型黑色背景。面对繁忙异常的主路，店面的深色流线背景和多彩标识散发出吸人眼球的具有流行元素的品牌形象。右侧的大片玻璃墙面将店内那些有趣的流行相关设计清晰地呈现在世人面前。

概念设计：DAAA & LOGIX 创意团队　建成时间：2014 年 12 月　面积：500 平方米　预算：250,000 欧元

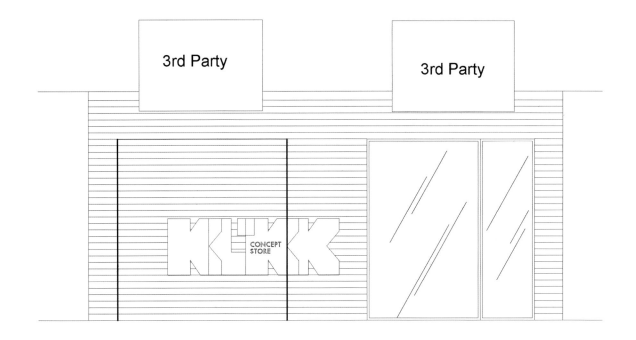

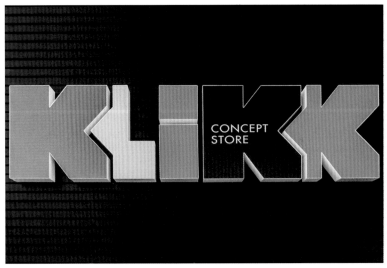

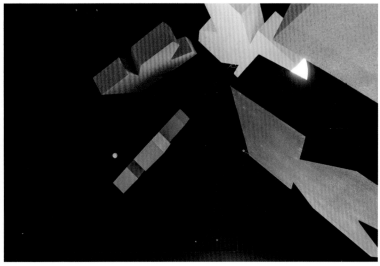

MODE:LINA ARCHITEKCI

FIESTA DEL VINO WINE BAR

Górczyn, Poland

"聚会之酒"红酒店

mode:lina architekci 建筑事务所 / 波兰，格尔兹恩

The owners wanted to expose the shop window and create a stylish character for the place that ultimately would redefine it.

The architects finished off the elevation with pallets, which they painted white and black. The building's form was simplified and was given a new skin. The designers exposed the steel logo of Fiesta del Vino and granted a new feel to the surroundings. The building is approachable, atmospheric and visible. Fiesta del Vino outstands in its location also thanks to the characteristic totem placed at the ramp.

Project Team: Paweł Garus, Jerzy Woźniak, Kinga Kin, Agnieszka Owsiany　Area: 120m²
Photographer: Marcin Ratajczak, Avantgarde Studios, Patryk Lewiński

委托方希望通过店面设计充分展示橱窗，同时打造店面的时尚特色，进而实现品牌形象的重新定义。

建筑师在外墙设计中使用了黑白配色。建筑形式得到简化，焕然一新。设计师选择充分展示店铺的钢质标识，为周围环境赋予新面貌。新店面亲切友好、感染力和可视性强。克服了周围环境的限制，脱颖而出。

设计团队：帕维尔·盖拉斯，耶日·沃兹尼亚克，金卡·金，阿格涅什卡·奥斯尼　面积：120平方米　摄影：马尔钦·拉塔扎克，Avantgarde Studios 工作室，帕特里克·雷文斯基

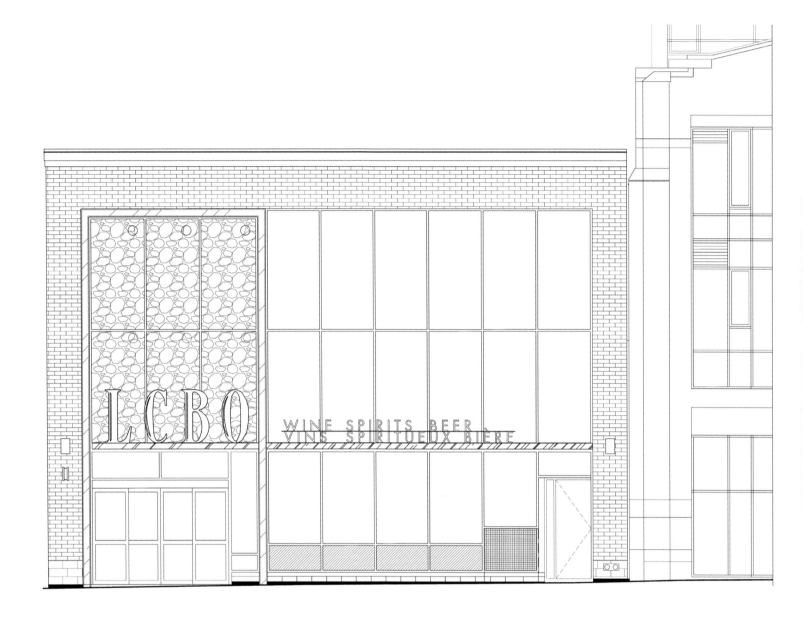

II BY IV DESIGN

L.C.B.O. URBAN INFILL STORE

Toronto, Canada

L.C.B.O. 城市葡萄酒店

II by IV设计公司 / 加拿大，多伦多

Located in the heart of a high-density, urban neighbourhood the Urban Infill Store is both modern in layout and functionality; however, possesses aesthetics that suggest hands-on craftsmanship and a feeling of old world charm. Design decisions were influenced by the understanding that urban customers want more than just convenience when shopping, they also want a selection of higher-end product and an 'experience' or 'retail-theatre'. Working closely with the client, the design team was responsible for creating a complete concept through construction package that included the exterior, as well as, interior design, graphics and fixtures programme.

On the exterior, a strong street presence is created by utilizing the full double storey façade of the building; defined by dark bronze mullions, Gabions (wire cages filled with rocks) are stacked above the entrance and a back-lit oversized graphic depicts idyllic, rolling hills. The effect is dramatic with the intention of lending a sense of history and context by referencing building materials and imagery typically associated with wineries.

Area: 8,600 square feet – 5,000 square feet of display space
Completion Date: June 2014

城市葡萄酒店位于城市里人口密集区的中心，无论布局还是功能层面都十分现代；同时它的外观设计又呈现出卓越的手工艺技术和复古的温馨氛围。城市顾客群体在消费过程中不仅要求便捷，也希望接触到大量更高端的产品，获得更"大气"的购物体验，这些信息直接影响了设计中的决策。设计团队与委托方密切合作，利用包括室外设计、室内设计、平面设计和家具设计在内的施工套餐实现完整的设计概念。

设计师在建筑外部使用两层楼高度的全外墙设计，在街景中打造出较强的存在感；在深色青铜竖框范围内，石笼（装满岩石的铁丝笼子）堆叠在入口上方，有背光装置的超大平面图案描绘出连绵起伏的田园山丘。建筑材料的选择以及葡萄酒酒庄的照片为店面增添了历史感。

面积：5,000~8,600 平方英尺的展示空间　建成时间：2014 年 6 月

JANNINA CABAL ARQUITECTOS

INTEGRAL ILLUMINATION COMMERCIAL BUILDING

Samborondon, Ecuador

整体照明商业大楼

约阿尼纳·卡瓦尔建筑师事务所 / 厄瓜多尔，圣波隆顿

Integral Illumination Commercial Building, the biggest light store in Samborondon, was developed in two diamond-shaped plots, while maintaining the exterior structure of the existing house (columns, walls and pitched roof) in one of the plots and proposing a new structure for the expansion of the building on the second plot.

The project is clearly defined by the merging of two different elements. On the left side, the existing structure wrapped up in a tensile fabric with a geometric-based pattern decomposed into irregular triangles. The facade structure made of a mixed-metal structure of steel and aluminium tubes wrapped in a medium grey tensile fabric developed specifically for outdoor use only. While on the right side a clean-cut double-storey glass prism is raised up on an exposed metal structure. The perception of the building varies completely from day to night. During the day the tensile fabric facade is a visual element which does not compete with the glass cube, creating a homogeneous structure altogether. At night, RGB coloured led reflectors brings the building to life highlighting the metal structure and accentuating the three dimensional effect of the panels in a game of lights and shadows.

Design Team: Jannina Cabal, Alejandra Lopez, Marco Sosa, Katty Cuenca   Area: 1,180 square meter   Photographer: Sebastian Crespo, Juan Alberto Andrade

整体照明商业大楼是圣波隆顿最大的商店，所在的场地呈两个菱形。项目要求在其中一个场地上保留原有的外部结构（立柱、墙壁和斜面屋顶），为另一个场地提出扩建新方案。

项目在两个不同元素的融合中获得清晰的界定。设计师在场地左侧的原有建筑上增加了有拉伸性的织物，织物上呈现解构成不规则三角形的几何图案。项目中还为外墙结构特别定制开发了钢铁和铝管混合金属结构，外部包裹中灰色拉伸织物，适合室外使用。场地右侧则是以金属框架为基础的棱角鲜明的双层玻璃结构。项目的外观在一天之中的不同时段不断变化。白天，外墙表面的拉伸织物的视觉效果不如玻璃结构耀眼，项目整体效果均衡。夜幕降临，建筑在红绿蓝三色LED反射物的衬托下迸发活力，突出金属结构，利用光与影的魔法强调外墙板材的三维立体效果。

设计团队：约阿尼纳·卡瓦尔，亚力杭德拉·洛佩兹，马可·索萨，凯蒂·昆卡　面积：1,180平方米　摄影：塞巴斯蒂安·克雷斯波，胡安阿尔贝托·安德拉德

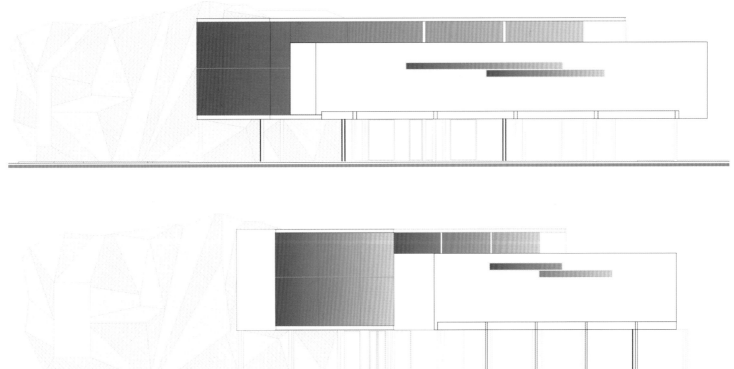

ARNAU VERGÉS TEJERO

A RELAXATION CONCEPT SHOP

Catalonia, Spain

放松概念店改造项目

阿尔瑙·维杰斯·德杰罗 / 西班牙，加泰罗尼亚

The store logo is an abtrast image of a tree, which expresses the attitude and philosophy of the brand and at the same time responds to the wooden elements inside. A brown colour scheme is choosen for the store facade, as it is neutrally welcoming, simple and pleasant for the eye. Logo elements can also be found on the side of columns beside the display window in order to interest passers-by.

The floor employs consistent stone material used by the whole building without special treatment. Moreover, brown decorative elements go perfectly well with the stone tone.

Designer: Arnau Vergés Tejero

店铺标识是一棵抽象的树，表达了整个店铺的风格与意境，与店内的原木装修材料也是一种呼应。外观装饰选用咖啡色做主色调，咖啡色属于大地色系，中性暖色色调，它优雅、朴素、庄重而不失雅致，是一种比较含蓄的颜色。橱窗旁边的柱子侧面也放置了标识，为了让过往的行人都能够注意到这个店铺。

楼面是整栋建筑统一的石材铺装，没有做特殊的处理。选用的咖啡色装饰与石材本身的颜色也非常匹配。

设计师：阿尔瑙·维杰斯·德杰罗

239 • 居家生活

MOMENT

LIVING MOTIF

Tokyo, Japan

生活主题馆

MOMENT 设计公司 / 日本，东京

As people's lifestyle keeps changing, a flexible shopfront can propose freshness to the customers. It is the freshness the design team attempted to present in this remodeling project. In this project, MOMENT designed the 'blank space' along the façade, which can change the appearance depending on the strategy to display freshness. It can be a retail space or a big showcase, tickling the curiosity of customers. A pioneer of design items store, 'Living Motif' is such a supple space, not affected by the trend. The team created a space as 'event space' mentioned above and a 'display area' on the first floor. Inside the event space along the façade, sometimes an exhibition and workshop is held, sometimes experimental lifestyle proposals are shown. Passers-by would also enjoy the event space even from the outside.

Completion Date: March 2015 Designers: Hisaaki Hirawata, Tomohiro Watabe Area: 1,200m² Photographer: Fumio Araki

我们的生活方式不断改变,灵活空间设计可以为顾客带来新鲜感。本案中,MOMENT设计公司打造了沿建筑外墙的"空白空间"。它可以根据展示策略改变外观。既可以是零售空间,也可以做大型展示空间,激发消费者的好奇心。作为先进的设计产品商店,"生活主题"馆不被潮流左右。设计师希望在这个翻新项目中体现新鲜感,紧挨外墙的特别空间有时会用来举办展览和研讨会,有时用来展示实验性的生活方式理念。路过的行人从外面也可以观赏到部分活动内容。

站在二楼,可以将曾经被玻璃覆盖的外墙尽收眼底,对店面的大小有一个直观的认识。突出店铺标识,地面被粉刷成黑色。

建成时间:2015年3月　设计师:平渡久明　面积:1,200平方米
摄影:荒喜文雄

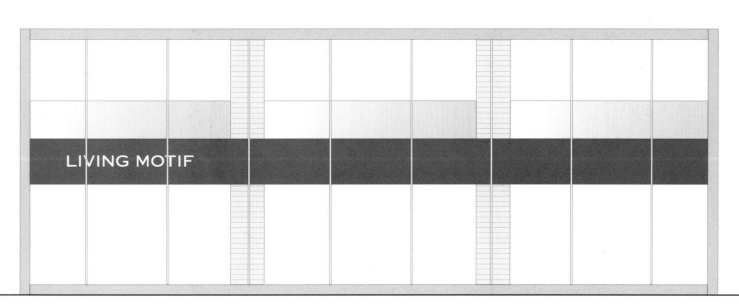

TRIPTYQUE ARCHITECTURE

TOG

Sao Paulo, Brazil

TOG 家具旗舰店
Triptyque 建筑事务所 / 巴西，圣保罗

The Flagship was designed as a mixed use place of sociability where you can buy, eat, drink, dance, read, create, spend time; a 21st century version of the Andy Warhol Factory. In TOG furniture, the space is fully customizable. Conceived as a multi-purpose, living space, the flagship store will also function as a major conviviality and sociable hub.

To anticipate these and others potential uses, the Triptyque agency neutralized this old games club's facade from the 80s in painting in white. There is a contrast between the surrounding walls and the totally new ground. The lights are suspended from the concrete beams visible on the ceiling through glass windows.

TOG 家具旗舰店是一个具备社交性的综合功能空间，消费者可以在这里购物、餐饮、跳舞、阅读、创造，度过一段美好的时光；是21世纪的安迪·沃霍尔工厂。这里提供全方位的家具定制服务。作为一个多功能生活空间，这间旗舰店同时也是一处主要的娱乐和社交会所。

为了探索这些预定功能以及其他潜在用途，Triptyque 建筑事务所将这个20世纪80年代曾经的游戏俱乐部粉刷成白色。这与周围建筑的墙壁以及全新的地面形成鲜明对比。照明装置悬垂在混凝土房梁上，透过玻璃窗可以看得一清二楚。

Completion Date: 2015 Total Area: 2,108.40 m² Design Team: Greg Bousquet, Carolina Bueno, Guillaume Sibaud, Olivier Raffaelli Project Manager: Alfredo Luvison Photographer: Ricardo Bassetti

建成时间：2015年 总面积：2,108.40平方米 设计团队：格雷格·布斯凯，卡罗琳娜·布埃诺，吉拉姆·席巴德，奥利维尔·拉法埃利 项目经理：阿尔弗雷多·鲁维真 摄影：里卡多·巴塞蒂

DALZIEL&POW
WYEVALE

Shrewsbury, UK

Wyevale 园艺中心

达尔齐尔与鲍设计公司 / 英国，什鲁斯伯里

Inspiration and hospitality are at the heart of Dalziel and Pow's new store design for Wyevale Garden Center, creating a complete lifestyle destination. Built from the ground up, the store in Shrewsbury features a dedicated education and service hub, The Greenhouse, where customers can browse a library of garden-related books, attend talks and watch demonstrations.

Two-thirds of the shopfacade is dedicated to home and lifestyle product, while cross-merchandized VM spots and room sets provide seasonal inspiration. With ambient pendant lights and spotlights, the overall feel is closer to a fashion or homewares store than a traditional warehouse-style garden center.

Signposting brand name acts as natural wayfinding aids to navigate shoppers around the 10,000sqm space. The brand name and the Garden Entrance both have a distinctive feel with different finishes and graphics, while a number of concessions are presented as standalone boutiques with their own entrances and windows.

Outside, more inspiration is delivered via a show garden displaying four different styles of garden that customers could achieve. These are joined by unique feature attractions including a sculptural metal and LED tree, and the world's largest spade.

Area: 10,000m² Completion Date: 2014

在达尔齐尔与鲍设计公司为Wyevale园艺中心打造的新店面中，灵感与热情是设计的核心方针，旨在打造一个全面的生活方式体验场所。位于什鲁斯伯里的这家园艺中心主营专业的教育与服务中心。人们可以在这个名为"温室"的地方浏览园艺相关图书，参加研讨会，观看展览。

店面外墙的三分之二都用于展示家居和生活类产品，同时交叉商品区和样板间提供季节性装饰灵感。环绕的侧灯和聚光灯营造出的氛围更像是时尚商店和家居用品店，而不是传统的仓储型园艺中心。

店内品牌的广告指示牌发挥导视作用，帮助消费者在10,000平方米的偌大空间里寻找方向。品牌名称和园艺中心的大门采用多样的质感和平面设计，带来个性鲜明的感觉。场地内的一系列独立精品店则有各自独立的入口和橱窗。

室外的展示花园呈现消费者可以实现的四个不同风格，进一步提供园艺灵感。包括金属雕塑、LED树以及全球最大的铁锹在内的特色景观则为整个店面增添了不少趣味。

面积：10,000平方米　建成时间：2014年

KAMITOPEN ARCHITECTURE-DESIGN OFFICE CO.,LTD.

ANTHEM BEAUTY SALON

Hitachinaka-city, Japan

颂歌发廊

KAMITOPEN 建筑设计事务所 / 日本，茨城县

Since hair salons don't sell items that are tangible, people come in with a mixed feeling of expectation, nervousness and anxiety. The designer created a salon where the stylist can ease the nervousness and anxiety of the customer by opening the door of a mirror as if he/she is opening the door of the customer's heart.

Moreover, among numbers of businesses, hair salon is one of the few businesses that don't have a fixed image. Therefore, the colour that is used the most in hair salons is 'white'. The designer thus thought of making a 'white space' which can change into all kinds of images. Customers feel colours before opening doors.

Designer: Masahiro Yoshida/Jin Hatanaka/Riyo-Tsuhata, Kamitopen Architecture-Design Office co.,ltd. Area: 260.67m² Photographer: Keisuke Miyamoto

由于发廊提供的不是实体产品，人们进店时会带有混合了期待、紧张和焦虑的情绪。在设计师打造的这间发廊里，造型师可以通过打开一门，展示门后的镜子缓解顾客的紧张和焦虑感，似乎打开了通往顾客心灵的大门。

而且，发廊是为数不多的没有固定形象的商业模式。因此，大部分发廊都会在店面设计中用到"白色"。设计师因此想到打造一个"白色空间"，使得顾客可以在开门前感受到鲜明的色彩。

面积：260.6平方米　设计师与设计公司：吉田雅宏，畑中仁，津秦世，KAMITOPEN 建筑设计事务所，吉田雅宏建筑设计事务所　摄影：启介宫基

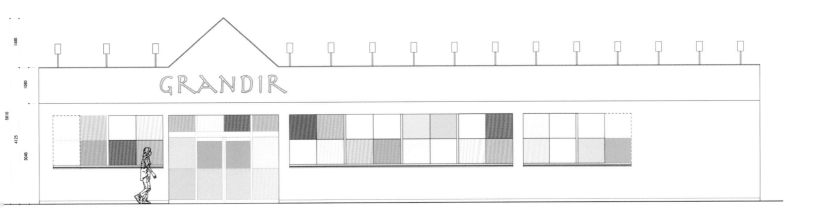

JASMINE LEE

THE BEAUTY CANDY APOTHECARY

Singapore

美人糖果药妆店

李茉莉 / 新加坡

The Beauty Candy Apothecary is a lifestyle concept store that brings together a curated selection of the best and up-and-coming beauty products and lifestyle accessories from around the world.

Inspired by the small boutiques in New York's Soho, London's Notting Hill, and Tokyo's Omotesando, the store sells a wide variety of hand picked products, many of which are available in Singapore for the first time.

With that in mind, the designer created a pure, clean look for Beauty Candy, drawing inspiration from typography of traditional apothecary trademarks to reflect the quality of products they carry. The minimalist white colour scheme creates a natural and comfortable ambiance. This is also employed in the entrance and interior, to achieve a refreshing elegance. Green plants with white entrance door add elegance and personality to the whole design. Mixed with decoration of graphics, the brand logo is pleasant to the eye and also displays an attitude of originality.

Completion Date: 2014

美人糖果药妆店是一间概念生活馆，汇集了全球最优质、最先进的美容产品和生活饰品。

店面设计灵感来自纽约Soho，伦敦诺丁山，东京表参道的精品店，店内出售种类繁多的精选产品。其中很多商品都是首次进入新加坡。

为了充分反映店面的特点，设计师从传统药妆店的商标设计中汲取灵感，打造了纯净、整洁的店面方案，体现店内商品的优秀品质。简约的白色调营造出自然、舒适的氛围。入口和室内空间采用了同样的设计，清新而典雅。绿植配合白色大门使整个设计的优雅与个性得到强化。品牌标识混用了装饰性图样，不仅美观大方，也展现出特立独行的品牌精神。

建成时间：2014年

NENDO

BEAUTY LIBRARY

Tokyo, Japan

美丽图书馆

Nendo 设计公司 / 日本，东京

Beauty products are displayed on rows of shelves arranged like a library inside this Tokyo cosmetics store by Japanese design studio Nendo. The studio, led by designer Oki Sato, created the minimal interior for organic cosmetics company Nature's Way.

The store is located on an alley behind Aoyama Street – Tokyo's premier shopping destination, where some of the world's most famous architects have designed flagship stores for luxury fashion brands. The comparatively small Beauty Library boutique occupies the ground floor of an unassuming building and has a fully glazed shopfront. Colourfully packaged products are arranged on floor-to-ceiling wooden shelving units, which are positioned in a line down the middle of the space – like book stacks in a library.

Photographer: Takumi Ota

在东京这间由 Nendo 设计公司打造的化妆品商店里，美容产品像在图书馆里一样，放在一排排架子上展示。Nendo 设计公司由设计师佐藤大木领衔，曾经为有机化妆品品牌"Nature's Way"打造了简约的室内设计。

本案位于青山街背后的小巷里，青山街是东京有名的购物区，在这里可以找到世界著名建筑师为奢华时尚品牌打造的旗舰店。相比之下，店面较小的"美丽图书馆"精品店位于一栋不起眼的建筑一楼，采用全玻璃外墙。包装鲜艳的产品在满满的木质货架上排列，让人联想到图书馆内的书架，十分有趣。

摄影：大田拓实

YOMA DESIGN

PRIM4 HAIR SALON

Taipei City, China

PRIM4 美发沙龙

YOMA DESIGN 设计工作室 / 中国，台北

The conception of the appearance design came from the combination of 'natural' signboard and interior space. The designers utilize lots of mountain element inside PRIM4, including mountain-shaped mirror and mountain-shaped entrance. Because the shape of mountain looks like home to them. They hope everyone in this store will feel like home, comfortable and relaxing. Large French window is a bridge for connection between outdoor and indoor. Door handle is made of water pipe, arranged in the shape of 'Four', which was made by YOMA given to PRIM4 as a souvenir. The signboard with grass material background and light design, given environment a humble greetings.

Area: 98.0m² Completion Date: 2014 Photographer: Figure x Lee Kuo-Min Studio

本案的外观设计从"天然"标识和室内空间的结合中获取灵感。设计师在室内空间使用了大量的山形元素，包括山形镜子和山形大门等。设计师希望进店的人能和他们一样，在山形元素中找到归属感、舒适感和放松的感觉。大扇的落地窗连接室内与室外空间，门把手是制成"4"形的水管，也是设计师为PRIM4美发沙龙特别制作的。标识板的草质背景和灯光设计，营造低调氛围。

面积：98.0平方米　建成时间：2014年　摄影：Figure x Lee Kuo-Min Studio 摄影工作室

MATT FAJKUS ARCHITECTURE

HILL COUNTRY APOTHECARY

Lakeway, Texas, USA

希尔乡村药店

马特·法库斯建筑事务所 / 美国，得克萨斯州，雷克维

The Apothecary includes a full-service retail pharmacy with products displayed on steel, aluminum, and poplar wood super graphic shelf that runs the full height and length of the glass storefront.

The super graphic combines the pharmacy's needs for clear and bold branding as well as ease of access to the products on display. The super graphic also manages direct southern light, adequately and naturally lighting the pharmacy interior while protecting sensitive items on the shelves.

Completion Date: 2014 Project Manager: Ian M. Ellis Design Team: Ian M. Ellis, David Birt, Matt Fajkus, AIA Photography: Cardinal Photo TX, Bryant Hill

本案中的药店提供全面的药品零售服务，由钢材、铝材和杨木制成的超平面风格货架占据整面玻璃外墙，用于商品展示。

超平面风格设计结合了药房对清晰大胆的品牌形象以及对货品展示的要求。它不仅能阻隔南部地区灼热的阳光，以充分而自然的方式为室内提供照明，同时也对货架上的敏感物品提供保护。

建成时间：2014年 项目经理：伊恩·M·埃利斯 设计团队：伊恩·M·埃利斯；大卫·伯特；马特·法库斯，美国建筑师协会会员 摄影：Cardinal Photo TX 公司，布莱恩特·希尔

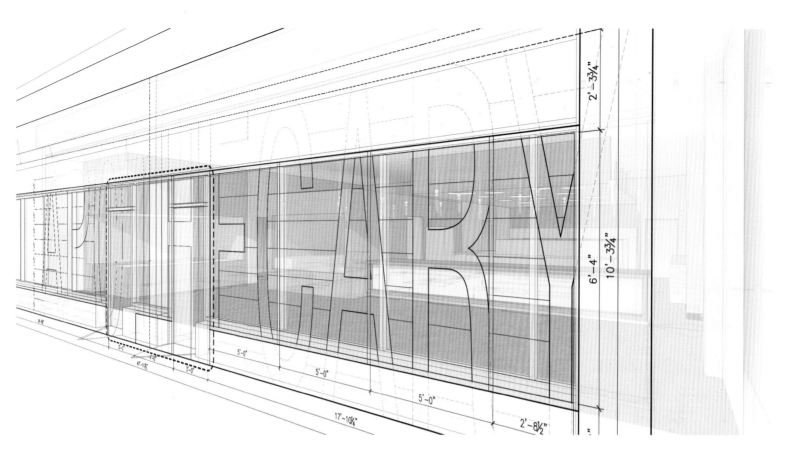

259 • 居家生活

SOPRATUTTO DESIGN TEAM

THE BOX PHARMACY

Athens, Greece

"盒子"药房

SOPRATUTTO 设计团队 / 希腊，雅典

The 'box pharmacy' is one of the most modern pharmacies and is designed to meet functional, ergonomic and commercial requirements of the owner. Interesting design also shows the facade of the pharmacy. The corner building is located in a busy intersection because of the abundance of inscription sand colours to neighbouring buildings could not be visually seen by passers-by. The both faces were coated with metal panels with specific laser cutting and painted with 3 colours as a result of three-dimensional and more reminiscent patch work. The large holes in the panels allow light to enter and reduce noise from the central intersection.

"盒子"药房是最为现代化的药房之一，它的设计旨在满足委托方在功能、人体工程学和商业层面的需求。这也是一个趣味十足的药房设计。项目所在场地位于繁忙交汇处的一间街角建筑内，原本十分不起眼。如今外墙的两面都安装了激光切割的特制金属板材覆面，并涂刷了三种颜色，打造三维立体效果。板材上的大型孔洞透光，并能降低来自十字路口的噪声。

Completion Date: September 2014　Designer: Elena & Giota Koutsopoulou　Area: 230m²
Photographer: Panagiotis Kinopoulos

建成时间：2014年9月　设计师：艾琳娜·库特索波路，宙塔·库特索波路　面积：230平方米　摄影：帕纳约蒂斯·基诺索波洛斯

LUIS GONZALO ARIAS RECALDE

PHARMACY EL PUENTE, GRANADA

Andalucía, Spain

格拉纳达埃尔蓬特药房

路易斯·冈萨洛,阿里亚斯·里卡尔德 / 西班牙,安达卢西亚

A pharmacy moves to a new location, and this enables the upgrade of an existing commercial premise—a former warehouse for the storage of ham—that had remained unused for years.

An aspect that has changed in pharmacies is the storefront. The traditional static display of products has ceased to make sense. What needs to be shown now is the pharmacy itself; the space has become the product.

The commercial premise has a large facade that faces a roundabout on the outskirts of the city. The new facade strives to capture the attention of the drivers that pass by. In order to achieve this, its finishing is extended to the undersides of the zigzagging cantilever of the balconies above the premise. The edge of the cantilever is lined with green LED lights that highlight the geometry of the building. This zigzagging line is about 12 meters long and it enhances the visibility of the business — subconsciously, it also recalls the traditional green cross of the pharmacy.

Completion Date: July 2015

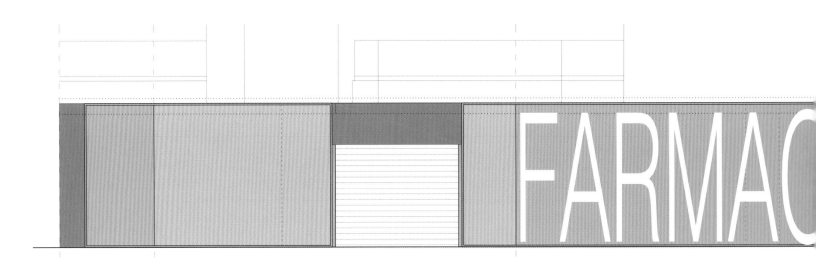

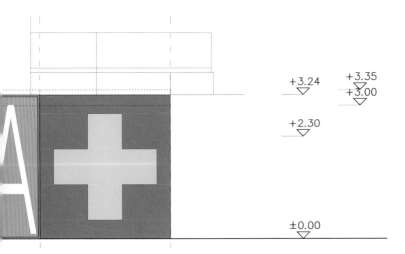

一间药房搬迁到新址，原有商业建筑因而得到升级。这里曾经是一间储藏火腿的仓库，多年无人使用。

药房发生变化的一个重要方面是它的店面。传统的静态产品展示已经不再有效，如今需要向顾客展示的是药房本身；这个空间就是推销的目标产品。

这处商业房产的外墙面积很大，正对着城郊的一个环岛。设计师希望新的外墙设计能够有效吸引路过司机的注意力。因而将外墙材料延伸到建筑上方锯齿状的阳台悬臂侧面。悬臂边缘与绿色 LED 照明装置平行，突出建筑的几何设计。锯齿状外沿长度约 12 米，它的存在突出了药房的醒目程度，也让人们在下意识中想起药房使用的传统绿色十字图样。

建成时间：2015 年 7 月

KDI CONTRACT

STELLATOU PHARMACY

Loutraki, Greece

Stellatou 药店

KDI CONTRACT 公司 / 希腊，路特奇

Source of inspiration in the design concept was water, derived from the vast natural springs and therapeutic spas of Loutraki. The element of water was utilized in the design process and was integrated in various forms, both in the design of the facade and the interior of the store. At the same time, large glass-panel openings, surrounding the pharmacy, were implemented in order to allow ample natural light flow, thus giving an open space feeling to the incoming clients.

The front façade shop-windows are transparent, without display shelves behind them, allowing an unobstructed optical communication with the interior of the pharmacy and providing ample window space for seasonal promotions. On the other hand, on the side façade shop-windows, custom-made display features were designed in order to exhibit the day-to-day products of the pharmacy. During the day, the façade of the pharmacy is the key visual element while at night, the illuminated interior of the pharmacy emerges, inviting the passer-by to visit the shop.

Completion Date: 2014　Project Director: Yiannis Kourtalis　Area: 125m²　Photographer: Giannis Gianelos

本案中设计理念的灵感源自路特奇储备巨大的天然泉水和水疗中的水元素。设计过程中运用到这一元素，并在外墙设计与室内设计中以多种形式得以呈现。环绕药店的大面积玻璃窗透入充足的自然光线，为进店的顾客营造开阔的空间感受。

店铺正面的橱窗是透明的，背后没有展示货架，与室内空间形成畅通无阻的视觉效果。宽敞的橱窗空间则用于季节性促销时使用。另外，店面另一侧的橱窗采用了定制设计，展示店内的日常产品。白天，药店的外墙是主要的视觉元素，而到了晚上，被灯光照得通亮的室内空间是吸引顾客的关键。

建成时间：2014 年　项目总监：扬尼斯·柯尔塔利斯　面积：125平方米　摄影：詹尼·吉安罗斯

EMMANUELLE MOUREAUX

SUGAMO SHINKIN BANK, NAKAAOKI BRANCH

Kawaguchi-shi, Japan

巢鸭信用银行第四分部

艾曼纽建筑设计公司 / 日本，川口市

The common request for all branches is to create a bank where people wish to stay longer and naturally feel to come back again. Nakaaoki branch is located on the corner of major intersection, where there is a frequent movement of cars, busses, bicycles, and people. Taking this unique location as a characteristic, the façade is designed to be rhythmical that changes expression as people see from different angles.

Colours appear in and out from the rhythmical repetition of cubes, dancing like musical notes playing rainbow melody. The façade is composed of cubes of four different depths. Colours are applied on the front or side of these cubes, so that the colours appear, disappear or overlap as the direction of view changes. Small elevated gardens are built inside 12 cubes, where the seasonal changes in nature are expressed by seasonal flowers like marigold, lavender, and growing trees such as olive tree. Gardens can be seen from the open space on the first floor, and from the financing section and cafeteria on the second floor. Sunlight is filtered through the foliage of elevated gardens on the South facing façade, providing a harmonious and warm atmosphere inside the bank. The interior finishes are settled and muted compared to the rhythmical movement of the façade. The colours, flowers and trees appear in and out from the repetition of floating cubes, playing rainbow melody. The melody spreads happiness and comfort to visitors and to the people in the local community.

Compltetion Date: June 2014 Area: 588.07m² Photographer: Daisuke Shima / Nacasa & Partners Exterior Finish: Aluminum plate fluoro-resin paint finish (cube), extrusion cement board fluoro-resin paint finish

巢鸭信用银行要求所有分支能够成为人们愿意停留更长时间并且自然而然地希望再次访问的银行。项目所在位置是主要交叉路口的角落，汽车、巴士、自行车和行人的活动十分密集。考虑到独特的位置条件，设计师为店面打造了一个从不同角度观察会规律变化的外墙设计。

外部深浅不一的立方体上跳跃着缤纷的色彩，就像演奏彩虹乐章的美妙音符。外墙上的这些立方体结构呈现4种深度，色彩出现在正方体的正面或侧面，随着视角的变化若隐若现，或交错在一起。12个立方体结构中是小型的架空花园，万寿菊、薰衣草和橄榄树等木本植物体现大自然的季节变化。人们可以在一楼的开阔空间，以及二楼的融资部门和自助餐厅欣赏到这些小花园。阳光穿过南侧架空花园中的繁茂枝叶，为银行打造出和谐温馨的内部氛围。与律动的外墙设计相比，室内空间较为静谧和平静。

色彩、花卉和绿植在彩虹般绚丽的小方体中散发着生命力，为客户与当地居民带来愉快、舒适的体验。

建成时间：2014年6月　面积：588.07平方米　摄影：岛大辅/Nacasa & Partners公司　外墙材料：氟树脂涂层铝板（立方体），氟树脂涂层挤塑水泥板

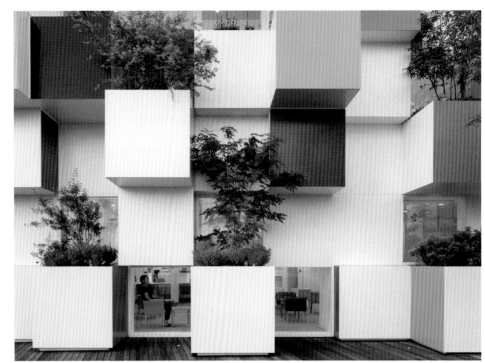

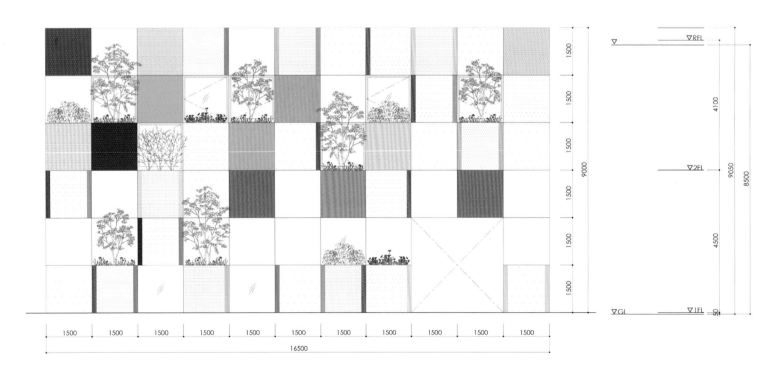

CAMILLE WALALA

THIRD DRAWER DOWN

Melbourne, Australia

Third Drawer Down 艺术设计商店
卡米尔·瓦拉拉／澳大利亚，墨尔本

Quirky art and design store Third Drawer Down has opened its second Melbourne store in the center of Prahran. Owner Abigail Crompton commissioned graphic artist Camille Walala to design and paint the exterior of the store, which was inspired by Nathalie du Pasquier's 1980s prints and the African Ndebele tribe, while lights and planters were provided by Kirsten Perry. The bright and colourful store façade features a standout graphic design in cobalt blues, turquoise, lemon yellow and monochrome, combining a series of geometric shapes and repeat chevron patterns. The store stocks design-oriented books and quirky homewares across furniture and crockery, as well as jewellery and accessories.

As for Third Drawer Down, owner Abigail Crompton specializes in bringing the cool, cultish American designers and brands down under (think Kiosk, Ben Medansky, Fredericks & Mae, Confettisystem) but her commissioned artist editions with the likes of Nathalie du Pasquier, David Shrigley, Ai Weiwei, and Louise Bourgeois have people wishing she'd open up a New York outpost.

Completion Date: February 2014

Third Drawer Down 是略显古怪的艺术与设计商店,已经在墨尔本开设了第二家店面。店主阿比盖尔·克朗普顿委托平面艺术家卡米尔·瓦拉拉设计并绘制了店面的外墙。卡米尔从 20 世纪 80 年代的娜塔莉·帕斯奎尔印花作品和非洲恩德贝勒部落中获得灵感,照明装置和花盆由克里斯汀·佩里提供。明亮多彩的店面外观采用钴蓝色、蓝绿色、柠檬黄色和黑白色构成的精美平面设计,结合一系列几何形状和重复的人字形花纹。店内存有设计类图书和奇异的家居用品,涉及从家具到陶器,再到珠宝和饰品在内的丰富内容。

建成时间:2014 年 2 月

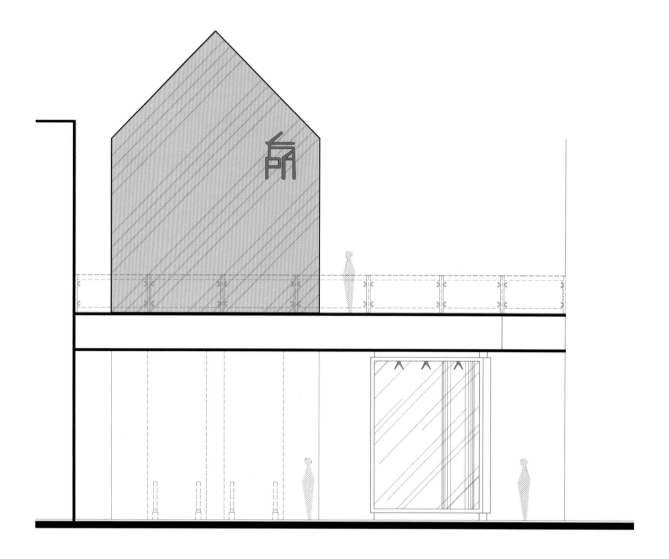

CHU CHIH KANG SPACE DESIGN
FANGSUO BOOKSTORE, CHENGDU

Chengdu, China

成都方所书店

朱志康空间规划 / 中国，成都

For Taiwan based designer Chu Chih Kang, bookstore design is more than a regular project but a hidden dream of 14 years. It is an honour in a designer's career to be able to create something beneficial to the society. The bookstore entrance resembles that of a house and employs the store's logo made of copper and iron, which will bear the prints of weather in time. The designer's concept of 'Scripture Library' was instantly accepted and carried out throughout the whole design process. At first, no one has any idea what the storefront should be like except that the client asked it to be related with Chengdu City. To achieve this, the design team carried out detailed investigation and discovered the connection between Daci Temple and Tang Sanzang, as well as the attitude Sichuan people hold against 'home' and 'chatting'. The Chinese people worked effortlessly for ancient wisdom in the search for Scriptures as long as thousand years ago. Scriptures and bookstores both are the source of great wisdom, therefore comes the idea of 'Scripture Library'. Books stand for unimaginable depth and thus bringing temple-like solemnity to the project.

Completion Date: February 2015　Leading Designer: Chu Chih Kang　Participating Designer: Jia Lu, Li Liuzhen, Li He　Area: 5,508m²　Materials: Copper, black iron, mosaic and concrete

书店设计在台湾设计师朱志康看来，不仅是一个项目，而是一个埋藏了14年的梦想。对于做设计的人，一辈子能够做一件对社会有贡献的项目，是非常荣耀的。入口仿照家屋造型，参考方所LOGO的设计，由金属铜与铁打造，会因天气变化而留下时间的痕迹。朱志康的"藏经阁"概念，一提出便获得认可，这一概念从未改变。初期，没有人知道成都方所该是什么样子，只有业主提出希望与成都有关。为此，团队做了详细的调查，发现了大慈寺与唐三藏的关联，包括四川人对"窝"和"摆"的生活态度。中国人早在千年前就为了寻找古老智慧而不辞劳苦，获取经书。而经书和书店都是智慧的宝藏，因此联想到藏经阁。书带有高深无法想象的寓意，所以便有了圣殿般的庄重。

完成时间：2015年2月　主要设计师：朱志康　参与设计师：贾璐、黎流针、黎合　面积：5,508平方米　主要材料：铜、黑铁、磨石子、混凝土

SFD, UNITED KINGDOM

INSTANT PRINT W1

London, UK

Instant Print "即时打印"冲印店

英国 SFD 设计公司 / 英国，伦敦

Instant Print approached SFD to redesign their print studio using a combination of vinyl and wood textures that could easily be altered to reflect the current season.

With reference from anamorphic design the front window frames the instore graphics, drawing your eye towards the newly formed collaborative working space at the rear of the studio.

The colour scheme of cyan, magenta, yellow and black reflects Instant Prints industry of printing whilst bringing life and colour into the studio. Similarly the vinyl aeroplanes on the window echo the daily use of paper at the workshop.

Completion Date: August 2015 Photographer: SFD, United Kingdom

Instant Print "即时打印"邀请 SFD 设计公司使用塑料和木纹元素为其重新设计印刷工作室。

以变形设计为参考，店内图案刚好位于正面橱窗的框架内，吸引人们关注工作室深处新建成的工作空间。

蓝绿色、洋红色、黄色和黑色的搭配组合一方面代表打印行业，同时又为工作室注入生机和活力。橱窗上的塑料飞机象征店内每天使用的纸品。

建成时间：2015 年 8 月 摄影：英国 SFD 设计公司

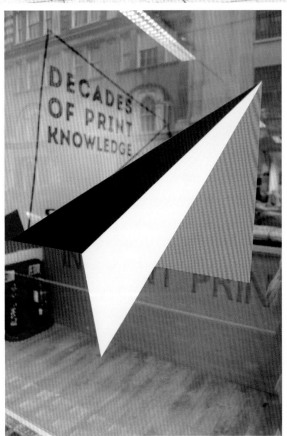

II BY IV DESIGN
6A ARCHITECTS

A
AAP (ASSOCIATED ARCHITECTS PARTNERSHIP)
ADAM WIERCINSKI ARCHITEKT
AKIRA KOYAMA + KEY OPERATION INC.
ANONYM
ARNAU ESTUDI D'ARQUITECTURA
ARNAU VERGÉS TEJERO
ARQUITETURA DAVID GUERRA
ATELIER ALTER

B
BARBARITOBANCEL ARCHITECTES
BAREA+PARTERS
BBC ARQUITECTOS
BEDMAR & SHI
BIASOL: DESIGN STUDIO
BRANDON BRANDING AGENCY
BROLLY DESIGN
BURO BLASÉ LOEFF

C
CAMILLE WALALA
CHIOCO DESIGN
CHRISTIAN LAHOUDE STUDIO
CHU CHIH KANG SPACE DESIGN
CLAUDIO SILVESTRIN ARCHITECTS
COLKITT&CO
CURIOSITY
CUT ARCHITECTURES

D
DAAA HAUS
DALZIEL & POW
DENYS & VON AREND

DPJ & PARTNERS, ARCHITECTURE

E
EMMANUELLE MOUREAUX
ESTÚDIO JACARANDÁ ARQUITETURA + DESIGN DE VAREJO
ESTUDIO VITALE
EXTERNAL REFERENCE ARCHITECTS

F
FMO ARCHITECTURE

G
GOSPLAN ARCHITECTS
GUAN INTERIOR DESIGN CO., LTD.

H
HANGZHOU GUAN INTERIOR DESIGN
HISANORI BAN KAZUMOTO TERASHIMA
HITZIG MILITELLO ARQUITECTOS
HOOBA DESIGN GROUP

I
IGNACIO CADENA
I LIKE DESIGN STUDIO
ING. ARCH LUKA KARIŽEK
ING. RADEK BLÁHA
IO STUDIO

J
JANNINA CABAL ARQUITECTOS
JASMINE LEE
JGV
JUN AOKI & ASSOCIATES

K
KAMITOPEN ARCHITECTURE DESIGN OFFICE
KDI CONTRACT

KENGO KUMA AND ASSOCIATES
KOIS ASSOCIATED ARCHITECTS

L
LABVERT
LINEHOUSE
LUIS GONZALO ARIAS RECALDE
LYCS ARCHITECTURE

M
MANUEL ATECA
MASAFUMI TASHIRO DESIGN ROOM
MASQUESPACIO
MATT FAJKUS ARCHITECTURE
MAURICE MENTJENS
MEI ARCHITECTS AND PLANNERS
MINAS KOSMIDIS (ARCHITECTURE IN CONCEPT)
MODE: LINA ARCHITEKCI
MOMENT
MORENO: MASEY
MOUSETRAP
MOVEDESIGN
MVRDV

N
NAN ARCHITECTS
NENDO
NERI & HU DESIGN AND RESEARCH OFFICE
NORDIC BROS. DESIGN COMMUNITY
NOT A NUMBER ARCHITECTS

O
OSCAR VIDAL STUDIO

P
PARTY/SPACE/DESIGN
PROCESS5 DESIGN

R
REIICHI IKEDA DESIGN

S
SFD, UNITED KINGDOM
SOPRATUTTO DESIGN TEAM
STUDIO DOTCOF
STUDIO RAMOPRIMO
SUITE ARQUITETOS
SUPERCAKE SRL

T
TACET CREATIONS
TALLER KEN
TALLER DAVID DANA ARQUITECTURA
TAI_TAI STUDIO
THE SWIMMING POOL STUDIO
TORII DESIGN OFFICE
TRIPTYQUE ARCHITECTURE

U
UUFIE

V
VAWDREY HOUSE
VINCENT LOEFF

W
WANG TAO
WATT INTERNATIONAL

Y
YONG-HWAN SHIN
YOMA DESIGN
YOSHIHIRO KATO ATELIER CO., LTD.
YOSHIHIRO YAMAMOTO
YOUSEF MADANAT ARCHITECTURAL STUDIO

图书在版编目（CIP）数据

商业店面设计. Ⅱ /（意）斯特凡诺·陶迪利诺编；张晨译. — 沈阳：辽宁科学技术出版社，2017.3

ISBN 978-7-5381-9879-9

Ⅰ.①商… Ⅱ.①斯… ②张… Ⅲ.①商店-室内装饰设计 Ⅳ.① TU247.2

中国版本图书馆 CIP 数据核字 (2016) 第 164710 号

出版发行：辽宁科学技术出版社
（地址：沈阳市和平区十一纬路 25 号 邮编：110003）
印 刷 者：恒美印务（广州）有限公司
经 销 者：各地新华书店
幅面尺寸：230mm×290mm
印　　张：17.5
字　　数：150 千字
出版时间：2017 年 3 月第 1 版
印刷时间：2017 年 3 月第 1 次印刷
责任编辑：杜丙旭　刘翰林
封面设计：周　洁
版式设计：周　洁
责任校对：周　文

书　　号：ISBN 978-7-5381-9879-9
定　　价：318.00 元

联系电话：024-23280367
邮购热线：024-23284502
http://www.lnkj.com.cn